THE BLACK BOX

ADVANCE PRAISE FOR THE BOOK

'This pioneering book is a must-read, not only for those interested in innovation policy in India, but also for those interested in understanding innovation and entrepreneurship in emerging economies more generally. A superb evaluation of policy in action, and the evidence on what works, what ought to but does not, but might yet do so in the future with some tweaking. R&D subsidies, tech transfer, incubators, patents, multinationals and human capital—the coverage is comprehensive and the analysis penetrating'—Ashish Arora, Rex D. Adams Professor of Business Administration, Duke University

'For a policymaker, the broad policy perspective provided by the book is very refreshing. It not only brings out interesting insights on the conventional innovation policy instruments but also highlights the emerging role of policies that support creation of a vibrant ecosystem for innovation-driven entrepreneurship in high-tech areas. The discussion on technology business incubators, public–private partnership models for start-up financing, the role of higher education and mission-driven approaches to support innovation, is especially useful'—K. VijayRaghavan, principal scientific adviser, Government of India

'It is rare to find someone who has experience in teaching, research and early-stage investing—the author draws from his unique vantage point at the junction of all three. The importance of various policy instruments in building a vibrant entrepreneurial ecosystem in India cannot be underestimated. This book provides useful insights on the efficacy of policies that support incubators and accelerators in higher education institutions to foster innovation-driven start-ups. The author draws upon the learnings from his teaching and research at IIM Ahmedabad and CIIE, and also from the experience of Infuse Ventures, a fund created through a public–private–academia partnership'—Sanjeev Bikhchandani, entrepreneur–investor

'Over the last sixty years, an increasing body of research has enriched our understanding of innovation. Rakesh Basant combines his deep understanding of theoretical work in the field with empirical studies of innovation in India. The result is an invaluable book, one that is required reading for all those interested in innovation studies, Indian industry or economic policy. *The Black Box* is scholarly, engaging, lively and thought-provoking, all at the same time. A long-awaited book by one of our leading scholars in the field'—Naushad Forbes, co-chairman, Forbes Marshall

'*The Black Box* is a patient and insightful tutorial on innovation and public policy—a compendium that is a must for all policymakers and innovation scholars, especially in India. It brings together a deep understanding of how the building of technology capabilities is enabled by governments and firms, and unravels the mysteries of development through technological innovation. It draws our attention towards the new role of higher educational institutions and innovation-driven start-ups in building a vibrant socio-economic environment in India. This is a timely book when policies have to be redesigned more sharply in the wake of massive technological changes that are happening around the world'—Pankaj Chandra, vice chancellor, Ahmedabad University

INDIA'S BESTSELLING BUSINESS BOOKS SERIES

AHMEDABAD
BUSINESS BOOKS

THE BLACK BOX

Innovation and Public Policy in India

RAKESH BASANT

PENGUIN
BUSINESS

An imprint of Penguin Random House

PENGUIN BUSINESS

USA | Canada | UK | Ireland | Australia
New Zealand | India | South Africa | China | Singapore

Portfolio is part of the Penguin Random House group of companies
whose addresses can be found at global.penguinrandomhouse.com

Published by Penguin Random House India Pvt. Ltd
4th Floor, Capital Tower 1, MG Road,
Gurugram 122 002, Haryana, India

First published in Portfolio by Penguin Random House India 2021
This edition published in Penguin Business by Penguin Random House India 2024

Copyright © Rakesh Basant 2021

All rights reserved

10 9 8 7 6 5 4 3 2

The views and opinions expressed in this book are the author's own and the facts
are as reported by him which have been verified to the extent possible, and the
publishers are not in any way liable for the same.

ISBN 9780670090822

Typeset in Sabon by Manipal Technologies Limited, Manipal
Printed at Replika Press Pvt. Ltd, India

www.penguin.co.in

This is a legitimate digitally printed version of the book and therefore might not
have certain extra finishing on the cover.

For
my father, Iqbal Narain,
who continues to be my benchmark for a liberal academic,
and
my teacher, K.K. Subrahmanian,
who introduced me to the economics of technology

Contents

List of Abbreviations

Alternative Investment Funds	AIFs
Artificial Intelligence	AI
Asian Development Bank	ADB
Bank of India	BoI
Biotechnology Industry Partnership Programme	BIPP
Build-Own-Operate	BOO
Business Model Innovation	BMI
Centre of Excellence	CoE
Centre for Innovation, Incubation and Entrepreneurship	CIIE
Centre for Monitoring Indian Economy	CMIE
Corporate Social Responsibility	CSR
Council of Scientific Research	CSIR
Defence Advanced Research Projects Agency	DARPA
Department of Energy	DoE

Department of Industrial Policy and Promotion	DIPP
Department of Science and Technology	DST
Foreign Direct Investment	FDI
Foreign Exchange Management Act	FEMA
Fund-of-Funds Scheme	FFS
Global Innovation Index	GII
Government of India	GoI
Gross Fixed Capital Formation	GFCF
Higher Education Commission of India	HECI
Higher Education Institution	HEI
Human Resource Development	HRD
Indian Council of Social Science Research	ICSSR
Indian Institute of Management Ahmedabad	IIMA
Indian Institutes of Management	IIMs
Indian Institutes of Technology	IITs
Industrial Credit and Investment Corporation of India	ICICI
Information and Communications Technology	ICT
Institutes of Eminence	IoE
Institutes of National Importance	INI
Integrated Circuits	IC
Intellectual Property Rights	IPR
International Finance Corporation	IFC
Investor Advisory Committee	IAC
Jawaharlal Nehru National Solar Mission	JNNSM

Just-in-time	JIT
learning-by-doing, by-using, and by-interacting	DUI
Machine Learning	ML
Micro, Small and Medium Enterprises	MSME
Ministry of New and Renewable Energy	MNRE
National Action Plan on Climate Change	NAPCC
National Innovation System	NIS
National Institutes of Health	NIH
National Research Foundation	NRF
National Science Foundation	NSF
National Skill Development Corporation	NSDC
National Skill Development Fund	NSDF
Plant Breeder's Rights	PBR
Private Equity	PE
Products, Processes and Practices	3 Ps
Protection and Utilization of Public Funded Intellectual Property	PUPFIP
Public Funded Intellectual Property	PFIP
Public-Private Partnership	PPP
Public-Private-Academia Partnership	PPAP
Purchasing Power Parity	PPP
Quality Function Deployment	QFD
Renewable Seed	REseed
Research and Development	R&D
Science and Engineering	S&E

Science and Technology	S&T
Scientific and Technologically-based Innovation	STI
Seed Capital Assistance Facility	SCAF
Small and Medium-sized Enterprise	SME
Small Business Innovation Research Initiative	SBIRI
Small Industries Development Bank of India	SIDBI
Smart Power for Environmentally and Economically Sound Development	SPEED
Special Economic Zones	SEZs
State Investment Banks	SIBs
Technical Assistance	TA
Technology Business Incubators	TBIs
Technology Development Board	TDB
Technology Licensing Offices	TLOs
Total Factor Productivity	TFP
Total Quality Management	TQM
Trade-Related Aspects of Intellectual Property Rights	TRIPS
Union Bank of India	UBI
United Nations Environment Programme	UNEP
University Grants Commission	UGC
University-Industry Linkages	UILs
US Patent Office	USPTO
Venture Capital	VC
Vocational Education and Training	VET

Preface

The term *innovation* is very commonly used in everyday conversations. Given the increasing recognition that innovation can be a potential source of development, writing on the subject has proliferated, both by academics and non-academics. Whereas the term innovation is used very broadly in non-academic literature and interactions, its conceptualization among the academics varies according to their disciplinary backgrounds and perspectives. The situation is not very different when one looks at the domain of *innovation policy*, although non-academic literature on this subject is limited. In general, academic writings on these subjects are not easy to comprehend while other studies often lack conceptual clarity.

Among various disciplines, economists and management scholars have contributed a lot to innovation studies. Recent studies on innovation suggest that there is value in combining the insights from these two disciplines to explore the relationships between innovation and public policy. Besides, the emerging literature and policy focus on innovation-driven entrepreneurship has added another dimension to the innovation-public policy interface. This book brings together conceptual and empirical literature from diverse disciplinary and intellectual streams in an eclectic manner to analyse the linkages between innovation and public policy. It makes an effort to bring out the complexity of these linkages that the vast literature on the subject has brought out but keeps the analysis simple and accessible. To restrict the scope of analysis at a manageable level, the focus is on *technology* innovations and not *business model* or *organizational* innovations. Also, while the approach is eclectic, the economist in me tends to dominate.

The book has been in the making for a long time. The roots of this endeavour actually lie in an elective course that I jointly offered with Pankaj Chandra at IIM Ahmedabad in the late 1990s. The course, called Policy, Technology Management and Competitiveness, covered many of the issues that are at the core of this book. Teaching, writing and interacting with Pankaj refined many ideas for me over time and added tremendously to my understanding of the *black box* which the economists call the Firm. His insistence on connecting policy to what happens inside the firm, including the shop floor, has stayed with me and has helped me understand the concept of technology, yet another *black box*, much better.

While Pankaj exposed me to some of the critical *micro* dimensions of the innovation and public policy interface, teaching a variety of public policy-related courses with Sebastian Morris and interacting with him for research and other activities for more than twenty years, has helped me appreciate the *macro* aspects better. Sebastian has been very generous with sharing his insights on various topics and without realizing it, I have started to use some of his frameworks to address certain policy questions. Through him, I understood the macro-foundations of the East Asian and Chinese experience of building innovation capabilities better, especially the role of exchange rates and cost of capital.

His insights on the models of public-private partnerships (PPPs) have also been invaluable.

Two other courses that I taught at IIM Ahmedabad helped me delve deeper into innovation and policy-related issues. One was a course on Economics of Regulation and the other on Innovation and Technology Policy. While the former provided exposure to the micro-economic foundations of various models that are used to analyse regulatory options, the latter gave me the opportunity to read the literature on innovation, especially from the development economics and national innovation systems perspectives. Sunil Mani, who taught the course on Innovation and Technology Policy before me, generously shared his material with me, helping me catch up quickly with the emerging literature in this area. Sunil has continued to help in a variety of ways, even in the writing of this book.

A large part of my tenure at IIM Ahmedabad has been spent working very closely with the Centre for Innovation, Incubation and Entrepreneurship (CIIE) at the Institute. This experience has been vital for my understanding of innovation-driven entrepreneurship in India as I was able to closely observe the impact of policies on the ground. We not only implemented the government policy on incubators in an innovative manner, but my association with CIIE also gave me the opportunity to be a part of an interesting policy experiment of supporting innovation-driven start-ups in the clean energy domain, which provided valuable insights on innovation financing using the PPP model. A significant part of the book also explores the impact of policies on innovative entrepreneurial ventures and various models of incubation and funding. Writing on this subject would not have been possible if I was not a part of the Centre. My interactions with Kunal Upadhyay have been central to my understanding of the critical issues in this space. Besides, without his help and insights, I could not have been able to write the story of Infuse Ventures and draw insights that have broader relevance. Apart from Kunal, A. Arumugam, Supriya Sharma and Vibhor Dhanuka also provided useful comments on the chapter on Infuse Ventures.

Interactions with Janak Nabar have been very informative and enhanced my understanding of some new policy initiatives. The Centre for Technology, Innovation and Economic Research (CTIER) has been generous in sharing data that they put together for their handbooks.

I have had a long association with Pulak Mishra. He is my go-to person for data and new literature in areas of common interest. His commitment and motivation are contagious. My long-term collaboration with him has helped me better understand the firm-level data provided by the Centre for Monitoring Indian Economy (CMIE). Punyashlok Dwibedy's help in analysing innovation-related data from the World Bank Enterprise surveys has made the empirical base of the analysis undertaken in the book stronger.

Neha Jaiswal provided excellent research support. She not only helped me with data analysis but also identified new literature relevant to the issues covered in the book. She was the first one to read the entire manuscript and helped me fix various anomalies.

This book would not have been possible without the generous support given to me as JSW Chair Professor of Innovation and Public Policy. Most importantly, this support provided the impetus that was needed to write this book and get me out of hibernation. In the absence of this push, many fragments of work undertaken by me over the years for teaching and research would not have found a somewhat cohesive and (hopefully) coherent articulation.

Appreciation at home of what you do at work goes a long way in making one productive. The same has been the case with me. My wife, Jeemol, daughter, Angika, and son-in-law, Sudarsan, have provided that support not only through their appreciation but also with their own productivity. Being part of a larger 'academic' family has always helped me in various ways. My siblings, Pankaj, Madhulika and Anupama and my brother-in-law Sushil remain an inspiration as they strive to do their best in their own academic careers.

I cannot undermine the role of the academic ambience that IIM Ahmedabad provides. Several unnamed persons have directly or indirectly contributed to this book. Students, who asked meaningful and, at times, inconvenient questions, colleagues who came up with considerate and impromptu comments with pointers to relevant literature, and friends who helped enormously by just being there. Among friends, I would especially like to mention Amita, Dinesh, Alice, Sebastian and the Shuklas, whose homes have always been like my own.

The book kept getting delayed due to a variety of reasons. But the enormous patience of Milee Ashwarya of Penguin Random House finally carried it through. It was a pleasure to work with Saksham Garg as the editor and Ralph Rebello as copy editor. His insistence on making the writing more accessible has enhanced the readability of the book.

The title of this book is inspired by Nathan Rosenberg's book called *Inside the Black Box: Technology and Economics*, which I read during my formative years as an academic. It was critical for building my understanding about the processes of technological change. My interactions with Professor Rosenberg in the 1980s, through snail mail, were the high points of my PhD days and helped shape the main argument of my dissertation.

I am really thankful to all the persons mentioned above for their insights, support and understanding. My mother always told me to strive to do better by looking up to those who are better researchers, teachers and human beings. I am yet to imbibe that lesson fully but as I look up, I know I could have done better. Still, I sincerely hope that the book will help create a more informed debate on issues of innovation policy as India strives to find a prominent place in the fast-changing global arena.

Ahmedabad   Rakesh Basant

1

Overview
The Agenda of the Book

1.1. Introduction

In 2019, the Global Innovation Index (GII) ranked India at 52, a remarkable improvement from the rank in 2015 which was 81. While this is good news, the ranks of countries like Vietnam, Thailand, UAE and Malaysia were better than that of India, and nations like the Philippines and Mongolia were not far behind with ranks of 54 and 53 respectively. China was far ahead of India with a rank of 14 (Dutta, Lanvin and Wunsch-Vincent 2019). GII is not a conventional measure of innovation but seeks to capture the enabling environment and infrastructure for innovation as well as innovation-related activities and outcomes. Elements of the national economy that can potentially enable innovation activities include institutional environment (political, regulatory and business) and market conditions for credit, investment, trade and competition.* Consequently, the index indirectly captures

---

* See Dutta, Lanvin and Wunsch-Vincent (2019), Appendix I for details.

a variety of policies that affect the institutional environment and the market conditions in which innovation activities are undertaken. Moreover, by including all these dimensions, the index also assumes a specific relationship between these dimensions and innovative activity in the economy. Implicit in this measure, therefore, is a relationship between innovation and public policy. But a consensus on such a relationship is yet to emerge.

It is widely understood now that innovation is critical for the development and growth of an economy. The world over, governments have worked with a variety of policies to encourage innovative activity. The capacity to develop new technologies and bring innovations to the market is considered to be essential for any nation to be globally competitive. Significant differences have been observed in technological development across countries, industries within the same country and firms within the same industry. Part of these *differences in technological performance* are typically traced to *different policy environments* which influence firms' responses to given market conditions as policies can differ significantly across nations, industries and firms. The remaining part of the variation in performance can be due to the *differences in capability to develop and manage technology*. The acquisition of the ability to absorb and build on technologies forms the core of technological capability-building and technology management endeavours at the firm level. Over a period of time, changes in economic policies can change the incentive structures and thereby influence the firms' behaviour vis-à-vis their investment decisions to develop technological capabilities. Therefore, in a dynamic sense, technological capability of firms and their responses to the policy environment are intertwined.

The state in developed and developing countries has sought to formulate policies which encourage technological capability-building at the firm level. Underlying any such policy is an understanding of factors that influence technological strategies of enterprises that affect innovation outcomes. Significant research has gone into the analysis of the complex linkages between public policy and innovation. While this research has generated a lot of interesting insights, several gaps remain in our understanding of these linkages. Given the complexity of the processes at work and the lack of adequate information, there is no consensus about the factors and the direction of their impact on firms' technological activities.

The Indian economy is currently going through a major transition process. As in other policy domains, a variety of innovation-focused policy instruments have been introduced in recent years. After the initial rounds of policy reforms, which have shown significant impact on economic growth, the reform process seems to have slowed down, partly because of the fact that the emerging policy challenges are complex and require significantly more analysis, and also because the political consensus around the next phase of policy reforms is difficult to build. Nonetheless, a large variety of innovation policy-related experiments are under way in India. These experiments not only relate to conventional policy instruments but also in the non-conventional domains wherein the key idea is to create a vibrant ecosystem for innovative and entrepreneurial activities. The book focuses on the emerging innovation policy challenges and highlights a few key areas which might require immediate attention.

Conceptualizing and exploring linkages between public policy and innovation is a complex endeavour. In order to address this complexity and disentangle some of these linkages in an accessible manner, the book is organized around a few interrelated questions:

i. How does one define innovation?
ii. What activities undertaken by various economic entities are relevant for generating and acquiring knowledge for innovation?
iii. Which policies affect these activities and how?
iv. What are the new policy initiatives that have been tried in emerging economies like India in recent years and with what impact?
v. Given the answers to the first four questions, what policy initiatives would be appropriate to foster innovation in the Indian economy?

Anticipating the discussion in subsequent chapters, the rest of this introductory chapter provides a flavour of the discussion on each of these questions.

1.2. Defining Innovation

Several definitions of innovation exist. While we discuss various definitions, the book itself focuses on *technological innovation* as against *business model, organizational* or *marketing* innovations. Popular literature tends to use definitions that cut across the conventional *Invention-Innovation-Diffusion* categorization and often confuses us between invention and innovation. While the book focuses on innovation, which entails bringing an invention to the market, it is recognized that economic growth requires all three processes to work effectively. Consequently, it is argued that this three-dimensional categorization remains useful for defining innovation appropriately and thinking through policy choices. The policy dilemma between innovation and diffusion is well known and we explore this further in the Indian context.

The analysis in the book works with a broad conceptualization of innovation, which is *change in technology* applied in the market. Technology itself is defined as knowledge embodied in *products* made and in the *processes* and *practices* used to make the product. Consequently, we explore innovations in all the 3 Ps—product, process and practices—and to the extent possible, analyse the links between them. Moreover, innovation can be incremental (adaptive), building on the existing technologies or radical (disruptive), changing dramatically the current ways of doing things. In other words, changing the technology paradigm itself. Studies have shown that the cumulative impact of multiple small incremental innovations is often higher than the impact of radical innovations. Consequently, incremental innovations are relevant for policy.

The book focuses both on the *inputs* (e.g., R&D expenditures) and *outputs* (e.g., patents, new products) of innovation to explore how choices relating to innovation-related activities and their outcomes are affected by policy. A synthetic example (Box 1.1) provides a peep into how innovation activities and policies may be intertwined at the level of a firm. But there are other entities who participate in innovation-related activities and affect innovation outcomes.

1.3. Choices and Activities of Economic Entities: Inputs to Innovation

What are the economic entities whose choices impinge on innovation outcomes? The literature identifies *government,* that plays a key role in deciding policies, *firms (small and large)* that undertake innovation-related activities, *higher education institutions* (HEIs) that generate knowledge and disseminate it partly through skill formation, and *other public and private institutions/organizations* that either play a role in linking the government with performers of research (e.g., research associations) or other institutions that facilitate the generation, protection and transfer of knowledge (e.g., public laboratories, patent offices, technology transfer organizations). The linkages between these actors determine the nature of knowledge flows across these entities and play a critical role in innovation.

To innovate, economic entities can use a combination of activities that involve *creating* (*making*), *buying* and *imitating* knowledge. Making would involve R&D, design, testing, training, etc. Knowledge can be bought in embodied and disembodied forms through the *acquisition* of machines, components, raw materials or even firms and *licensing*. Technological efforts of others in the market create imitation potential (spillovers). Exploiting this potential through imitation can take the form of reverse engineering products, attracting employees of competitors, using consultants or analysing patents and inventing around them. The book suggests that these choices are complex and typically dependent on absorptive or innovation capacity of the economic entities involved, nature of innovation and so on.

1.4. Public Policies and Their Potential Impact: Competition and Contagion Effects

Firms' choices about their innovation activities are affected by a variety of factors including the competition they face, the extent to which they can protect their innovation from imitation and the availability of innovation-related knowledge from various sources (Box 1.1). Innovation can be a way to respond to competition, Intellectual Property (IP) protection may enhance the incentive to undertake innovation-related activities and availability of knowledge creates opportunities to buy as well as imitate, given the potential to learn from others. In this context, the book explores several interrelated questions. How do policies influence the choices or activities of firms and other entities? What implications do different policies have for the supply and demand of knowledge (innovation) and consequently these choices? How do policies change competition and contagion conditions and in turn affect the choices of economic entities?

Box 1.1
Shaping Technology Strategy of an Indian Firm: A Synthetic Example

Panacea, a pharmaceutical firm in India, was reviewing its technology-related investments in the year 2000. A significant part of its manufacturing equipment was procured domestically during the pre-liberalization phase as imported machines were very expensive due to high import duties. These machines needed to be replaced. Before 1991, the weak IP regime in India had led the company to focus on process-related R&D since product patents were not available in India and new drugs could be invented relatively easily for domestic markets. However, due to the impending change in the IP regime as India had signed the TRIPS agreement, Panacea had initiated some product-related R&D in the mid-1990s. It helped the company understand research outputs of other firms, both domestic and foreign. Severe restrictions on foreign technology licensing during the pre-1991 period also meant that Panacea had not licensed any foreign technology. As the senior management thought about the company's technology strategy, they had to take cognizance of the new reality, which among others had the following elements:

- Foreign markets were growing very rapidly for biosimilars and generics due to trade liberalization but these markets required higher quality. Exports can not only enhance quality but also facilitate other knowledge flows from global markets
- Achieving higher quality required good manufacturing facilities and at least some internal R&D for both processes and products
- Manufacturing equipment (embodied technology) could be more easily imported now due to lower duties and few import restrictions
- R&D expenditures provided significant tax benefits in the new policy regime
- Technology licensing policy had been liberalized, making purchase of disembodied technology easier and a firm can potentially combine such licensing with its own R&D efforts
- Inventing new drugs was significantly more difficult if not impossible in the new IP regime and often required more internal R&D effort

- IP protection of technologies developed by the firm was more important now than earlier for the product as well as process-related inventions
- Liberal policies of outward foreign investment made acquisition of foreign firms by Indian firms easier
- Due to liberalization of policies, competition in both domestic and foreign markets was on the rise and required a technology-driven response

As the senior managers thought about making, buying or imitating technologies for their future growth, they considered the new set of options that had become relevant for them to procure embodied and disembodied technologies, due to the changes in policies and the nature of market competition.

It is argued that public policies influence demand and supply of knowledge in the market apart from affecting the degrees of competition and contagion (spillovers of knowledge or the potential to learn from others). These in turn have an impact on strategic choices made by economic agents on knowledge generation and acquisition. Make-buy-imitate choices relating to innovation are affected by competition and contagion conditions. As mentioned, investment in innovation may take place to meet an increase in competition but the absence of 'monopoly rents' due to prevalent IP regimes (or presence of contagion) can be a disincentive to such investments. Typically, as various agents—firms, universities, research organizations and individuals—focus more on innovation-related activities, both the demand and supply of knowledge would increase. And the contagion effect (imitation potential) increases with larger pools (supply) of knowledge to learn from. The book argues that policies affect both competition and contagion conditions and the impact of these conditions on innovation is mediated by innovation and absorptive capacity of the economic agents.

Once activities that result in innovation are defined as above, a large number of policy initiatives can influence innovation outcomes. Analysing the complex interactions among these policies and their impact is probably the biggest challenge to understand the interface between public policy and innovation. The book tries to unravel these interfaces to the extent possible. To anticipate part of the discussion and illustrate the complexity, a few policy-innovation linkages are enumerated below to show how various policy instruments affect competition and contagion conditions faced by economic entities.

Industrial policy affects entry and exit barriers and hence, competition conditions. *Foreign Direct Investment* (FDI) *policy* influences both competition and contagion conditions as multinational corporations (MNCs) add to competition by their entry and also bring in new knowledge to create contagion potential as local firms come in contact with foreign firms. Among other things, domestic firms can observe the activities of MNCs, unbundle their products (to some extent even their processes and practices) and build a variety of linkages with them to learn from them. *Trade policy* affects access to embodied innovations through imports of machinery, etc., and competition through imports-based entry. It also has contagion potential as high technology intensity of imports can be a source of learning. Learning can also happen through exports as firms learn from demanding customers abroad and bring back new knowledge as they compete abroad.

Education and S&T policies help competition and contagion by generating new knowledge, creating an absorptive capacity to exploit spillover potential and attracting innovation-intensive investment, etc. *Intellectual Property Rights (IPR) policy* can affect competition and contagion conditions in a variety of ways. A strong IPR regime can create incentives to innovate due to monopoly rights (R&D, patenting) and by aiding the development of technology markets. A weaker regime makes imitation easier. *Technology licensing policy* affects extent, price and vintage of technology licensed, and transfer of knowledge, especially tacit, through training. Entry or competition through technology licensing and the pool of knowledge created through the transfer has implications for competition and contagion conditions. *Policy on standards* affects demand and supply side network economies and innovation to meet standards. *Policies affecting funding and cost of capital* can influence investments in new technology both through purchase and R&D. *Venture Capital* (VC) funding may help innovation-based entry by start-ups and debt funding for purchase of embodied innovation. *R&D subsidy and tax concessions* provide incentives to undertake R&D which in turn results in an increase in the knowledge pool (contagion potential) and the possibility of innovation. *Cluster policies* facilitate agglomeration and coexistence of competition and collaboration among cluster firms and institutions. Flows of knowledge happen in these agglomerations through various mechanisms including participation in global production and innovation networks by cluster firms.

Significant research has gone into the analysis of the complex linkages between public policy and innovation. As mentioned, while this research has generated a lot of interesting insights, it has also identified several gaps in our understanding of these linkages. This book is an attempt to pool together some of the ideas that academic research has highlighted on the linkages between technological innovation and public policy and identify the current challenges as well as opportunities for meaningfully exploring these linkages further, apart from generating some policy insights. While we use evidence from across the world, the empirical focus is essentially on India. We use both secondary data and case studies to explore the complex set of questions summarized in this chapter. It is also hoped that the discussion in the book will be useful both for managers and policymakers apart from providing some material and insights for those academics who wish to undertake innovation-related studies.

The rest of the book is divided into five chapters. The next chapter explores the concept of innovation and its measurement. Participants in the innovation process and the activities that contribute to innovation are discussed in detail. Chapter 3 focuses on different dimensions of innovation policy. It provides an overview of the public policy-innovation interface and discusses the mechanisms through which various policies impact innovation. Available evidence on policy-innovation linkages in different parts of the world is presented along with a discussion of challenges of empirically exploring such linkages. Chapter 4 focuses on innovation policy in India and its innovation inputs and outputs. The focus is on some new policy initiatives that have been tried in India in recent years, including those that support innovation-driven entrepreneurial ventures. Insofar as university-industry linkages and availability of early-stage venture investment are critical for creating an ecosystem for such enterprises, the chapter also discusses some recent developments in these areas. Chapter 5 extends the analyses contained in Chapter 4 by focusing on issues relating to financing innovation in the context of a policy experiment undertaken in India to support start-ups in the clean-tech space. The final chapter summarizes the key insights of the book and identifies policy initiatives that would be appropriate to foster innovation in the Indian economy. It also identifies a few potential areas of research in the area of innovation and public policy.

2

Innovation

Definition, Activities and Actors

2.1. Introduction

Innovation has been defined in a variety of ways.[*] As mentioned in Chapter 1, we focus on technological innovation[†] and take an economist's perspective on technological change that recognizes the utility of distinguishing between *invention*, *innovation* and *diffusion*. An invention can take different forms—a novel product, method or process. The wheel, printing press, electricity, telephone, personal computer, penicillin, automobile, calendar, vaccines and airplane are

---

[*] Baregheh, Rowley and Sambrook (2009) reviewed a large number of definitions of innovation from multiple disciplines and came up with the following: 'Innovation is the multi-stage process whereby organizations transform ideas into new/improved products, services or processes, in order to advance, compete and differentiate themselves successfully in their marketplace' (p. 1334). For more discussion on the definition from a multidisciplinary perspective see, Nair et al. (2015) and Crossan and Apaydin (2010).

[†] Innovations in marketing, business models, etc., are not our focus here.

all well-known product inventions. But underlying all these product inventions are many new processes and methods without which these product inventions may not have been a reality. When any *invention* is introduced in the market (or used), it becomes an *innovation* and as more and more economic entities adopt it, the *diffusion* process ensues.[*] It is critical to recognize that without the unfolding of the diffusion process, a technological innovation cannot have an impact on the economy and society. Therefore, policy needs to ensure that economic agents actively participate in *all* the three 'stages'—invention, innovation and diffusion—of technological change. While this distinction is critical for a variety of reasons (some of which are discussed below), it does not mean that technological change is a purely linear process. The feedback loops between these three stages of technological change and their importance in understanding the innovation process is well recognized in the innovation literature.[†] Some of these dimensions are discussed later and it will be argued that policymakers need to worry about certain trade-offs between the three stages as they evaluate policy initiatives to create incentives for agents to participate in these stages.

Technology has been defined as *knowledge* embodied in *products*, *processes* and *practices* (3 Ps) (Chandra 1995;

---

[*] This is often referred to as the Schumpeterian typology of technological change: 'invention (the generation of new ideas), innovation (the development of those ideas through to the first marketing of or use of a technology) and diffusion (the spread of technology across the potential market).' (Stoneman and Diederen 1994).

[†] The seminal work by Kline and Rosenberg (1986) initiated a lot of research on coupling and chain-linked models. The systems of innovation concept (based on interaction and learning), which is now used pervasively, has its roots in the coupling models.

Basant and Chandra 2002)*. This knowledge can be *embodied* in machines, raw materials and other inputs or *disembodied* in the form of a technology licence, a patent document or a technology manual. It can be *explicit* when it is codified in certain documents or *tacit*, embedded in the practices/ routines of an enterprise or in the minds of its workforce. *Change* in technology (or knowledge embedded in any of the 3 Ps) introduced in the market is *innovation*. For example, the telephone as a device has undergone significant changes over time. Both the functionality and the look and feel of the product have changed dramatically. With the advent of cell phones, the changes have been even more dramatic. The materials, methods and processes used for producing the phones have changed, apart from the mechanisms that make them work. Consequently, the underlying knowledge embedded in the product features and in the processes and practices used in producing the devices have all changed. This change in knowledge is innovation in all the 3 Ps. It needs to be recognized that while in the case of the phone, there have been significant innovations in all the 3 Ps, this may not be true in all cases. Enterprises or individuals as a result need not always focus on *all* the 3 Ps for innovation. A pharmaceutical company, for example, may focus its innovative efforts on changing the process and/or practice of an existing medicinal drug without changing its product characteristics. The resultant cost savings may make the firm more competitive.

Once innovation is seen as an introduction of *new* knowledge in the market and one views the innovation process

---

* For similar conceptualization, see Lipsey (2002). The perspective taken here is different from Lipsey's insofar as the concept of practices (which includes the knowledge regeneration process) is wider in scope than his 'organizational routines'.

from the perspective of an enterprise, it is useful to distinguish between *degrees of novelty* that the new knowledge entails. The literature typically classifies it as incremental (adaptive) or radical (disruptive).* Moreover, the knowledge may be (i) *new* to the *enterprise* under consideration; (ii) *new* to the *market* that the enterprise operates in; or (iii) *new* to the *world*. And as mentioned, the *new* knowledge can relate to *any* or *all* of the 3 Ps. The focus on incremental and new to the market vs new to the firm blurs the dichotomy between innovation and diffusion, as new to the firm may not be new to the market but is quite important from the perspective of diffusion. Besides, the complexity of the innovation-diffusion phenomenon increases once it is recognized that often, adaptation of new knowledge is required as a part of the diffusion process and adaptation often requires incremental innovation. Many studies have shown that as an innovation diffuses in new areas or contexts, neither the innovation nor the environment in which it is diffusing remains the same. Consequently, a strict separation between innovation and diffusion processes is often not observed in practice (Metcalfe 1985; Basant 1990). Moreover, since the cumulative impact of incremental innovations is often higher than the impact of radical innovations (Abernathy and Utterback 1978; Freeman and Perez 1988), it is especially relevant for policy.

The policy dilemma regarding the trade-offs between innovation and diffusion has been well known—policies that facilitate invention and innovation may constrain diffusion.†

---

* While these are widely used categories of innovation in the literature, several typologies of technological innovation exist which can be quite confusing. See Garcia and Calantone (2002) for a critical review.

† See, for example, Ordover (1991) and Stoneman and Diederen (1994).

For example, strong and exclusive intellectual property rights may create incentives for invention and innovation but may be detrimental to diffusion. But diffusion of technology is critical for enhancing productivity and competitiveness of a sector or economy. It has been suggested that, till recently, policy initiatives have not focused on the improvement of the diffusion process (Stoneman and Diederen 1994).

Next, we discuss the concept of innovation in detail which is followed by a description of activities that contribute to innovation and of participants in the innovation process. The defining of the purpose of innovation policy is to create circumstances in which various participants in the innovation process undertake activities that result in innovation. In the process we identify issues that policy needs to address in order to create an innovating economy.

2.2. Conceptualizing Innovation

As noted, there are multiple types of innovation. Innovation can take the form of introducing new *products* or new production methods (*processes*) or organizing production in new ways (*practices*). Utilization of new supply sources or materials may entail changes in one or more of the 3 Ps.* Recognition of the three types of innovation makes strategic sense as, given the availability of resources, organizations can make choices about which Ps or combinations thereof they need to focus on for enhancing their competitiveness in the market in different time frames. Such a recognition is

---

* In the same vein, exploiting of new markets, geographies, sectors or uses can result in modifications in one or all of the 3 Ps. There can also be improvements in the instruments or methods of doing innovation which can also be seen as innovations in practices.

also useful for policymakers as different policies may affect different Ps, creating trade-offs for policy choices. Moreover, *product* innovation for one enterprise can result in *process* innovation for another. For example, a new machine (product) developed by an industrial machinery firm can become a critical part of *process* innovation for another enterprise that *uses* this innovative machine for producing a product. This new machine can also help the user of the machine to develop a new product. Thus, if policy creates incentives to innovate for the industrial machinery firms and not for the users of their innovations, innovative activities might suffer in both sets of firms due to the input-output or product-process linkages. We will get back to this issue in Chapter 3.

2.2.1. *Product innovations*

When the word innovation gets mentioned, people often think about product innovation. This is so because as consumers they are exposed to product innovations more directly as users. At the 2016 Consumer Electronics Show, the electronics company LG introduced a new type of screen that is so flexible, one can roll it up like a newspaper. This innovative product solves the problem of portability—instead of using a large, unwieldy screen, people can show a video on a screen that they can fold up once they are done and put it in their bag! One can think of a variety of such wonderful new products but not all product innovations are so radical, disruptive and fanciful. Conceptually, one can categorize product innovation in three ways: (i) a new product that did not exist before, such as the Fitbit or Amazon's Kindle; (ii) an improvement in the performance of the existing product, such as an increase in the digital camera resolution of a cell phone; and (iii) a new feature in an existing product, such as power windows in a car. Both technological advancements and changes in customer requirements can provide an impetus for product innovations. Typically, innovations of type (ii) and (iii) are more common than the ones of type (i). And as mentioned, some of these innovations may not be new to the market but can be new to the enterprise.

Box 2.1
Products, Processes and Practices: The 3 Ps

Technological innovation entails change in knowledge used by an organization. This knowledge is embodied in the *products* (or service) produced by an enterprise and the *process* and *practices* used by it to produce, sell, deliver and support these products. While changes in products and processes are easier to comprehend, practices encompass a variety of activities, ranging from practices like *just in time* or *quality circles* (including certification), *use of Information and Communication Technologies (ICTs) in various systems* and systems that facilitate the *knowledge regeneration process* (including training). Defining the 3 Ps might be simpler in the context of manufacturing but for other sectors, boundaries may not be as clear. But the distinction is useful nonetheless, both from the strategic perspective of an enterprise as well as from the perspective of the policymaker (see text).

In the software sector, *products* are the applications developed by the enterprise, *process* comprises computers, programming software or tools used, telecommunications, storage devices, programmers and system designers, and *practices* include parsing rules, rules for organizing codes, debugging and test procedures and so on. Similarly, for the banking industry, different instruments of lending, investment schemes and various services provided to customers would be the *products*. *Processes* would include asset management software for investment banking, banking norms and procedures, forecasting systems, etc., while *practices* among others would consist of rules of operations, scheduling, staffing, designing quality in services, researching new investment options or developing new products.

Source: Based on Basant and Chandra (2002: p. 401).

2.2.2. *Process innovations*

Broadly, a process is the combination of facilities, skills and technologies used to produce, deliver and support a product or provide a service. Within these broad categories, there are several ways process innovation can take place. It can include changes in the equipment and technology used in manufacturing (including the software used in product design and development), improvement in the tools, techniques and software solutions used to help in supply chain and delivery systems, etc. One of the most famous examples of process innovation is Henry Ford's invention of the moving assembly line.* This process not only simplified vehicle assembly but shortened the time necessary to produce a single vehicle.

---

* In 1913, Henry Ford installed the first moving assembly line to mass-produce an entire automobile. According to some estimates,

While product innovation is often visible to customers, a change in process is typically only seen and valued internally within an enterprise. Competitors also value it as well and get to know about it through various mechanisms—information revealed in a process patent document, consultants, suppliers, collaborators, employees from competing firms and so on. Generally, changes in the process reduce costs of production more often than they drive an increase in revenue.

2.2.3. Practice/Organizational innovations

Practices are the 'grammar or the language necessary to manage the product-process combine' (Basant and Chandra 2002: p. 401). In the manufacturing context, just-in-time (JIT), *Kaizen* and *Kanban* are well-known practice-related innovations introduced by the Japanese enterprises and then adopted by firms across the world. Total Quality Management (TQM), a response by the US to the Japanese innovative manufacturing-related practices, also encompasses a set of customer-oriented practices that facilitate continuous improvement. Similarly, the Quality Function Deployment (QFD) method is an efficient mechanism to manage the product-process combine as it helps define customer requirements and convert them into detailed engineering specifications and plans to produce the products that fulfil those requirements. Many of these practice innovations take the form of new organizational routines that help produce, deliver and support a product or provide

---

this innovation reduced the time it took to assemble a car from more than twelve hours to two hours and thirty minutes. For some more details, see: https://www.history.com/this-day-in-history/fords-assembly-line-starts-rolling

a service in a more efficient manner. These practices may be used to sell and maintain the product the enterprise produces as well as methods used for accounting and customer service. Practices used by organizations for *knowledge regeneration processes* are critical as they define the learning or capability-building mechanisms and help the firm acquire a dynamic competitive advantage.

Business Model Innovation (BMI) is a buzzword these days and discussions on such innovations bring up names like Airbnb, Uber or Spotify. These are well-known examples of enterprises that have managed to disrupt age-old markets (hotel, taxi and music respectively) by tweaking or inverting their industry's traditional business model. While the definition of BMI is still evolving, it does not necessarily imply changes in the product or even in the production process. In its simplest form, a business model can be defined as a mechanism to create, deliver and appropriate value. Any change in these mechanisms can be viewed as BMI. Some studies view BMI as a complement to product, process and organizational innovations (Zott et al. 2011) while others suggest that it is a more holistic form of organizational innovation (Foss and Saebi 2017). Many features that are attributed to new business models can actually be subsumed under *practice* innovations, especially those that relate to innovative ways of delivering value.* These can include innovations in logistics, delivery, distribution methods, maintenance systems or organizational operations (purchases, accounting, data analytics) and even managerial practices. Similarly, marketing innovations (new or improved marketing methods) can also be part of BMI

---

* Foss and Saebi (2017) suggest that BMI may be undertaken to reduce costs, optimize processes, introduce new products or access new markets.

and *practice* innovations as well.* Like practice innovations, BMI can be undertaken both by start-ups as well as existing firms—both small and large.

Insofar as many new practices (as well as BMI)† have origins in the development of the ICT sector, the developments in these technologies and their adoption by organizations have provided significant impetus for innovations in practices and have also enhanced the efficacy of existing practices. In fact, these changes have also affected product as well as process innovations but broadly, the impact seems to be higher on practice and BMIs. ICT adoption has changed organizational practices in significant ways and the examples of BMI mentioned above (Airbnb, Uber or Spotify) ride on developments and adoption of ICT. Consequently, adoption of ICT is not only an innovation in itself; it can be leveraged for many new innovations, especially in the practice domain. Since access to ICT is influenced by public policy, state initiatives in this domain can be very critical for innovation. With the advent of machine learning, artificial intelligence and big data analytics, the importance of ICT access and use would continue to grow. We shall revert to this issue in Chapters 3 and 4.

---

* Business model innovation is not the focus here but the 3 Ps framework can capture many BMIs especially through the *practice* component. Hossain (2017), Foss and Saebi (2017) and Ramdani, Binsaif and Boukrami (2019) provide useful reviews of studies on BMI.

† Some studies actually describe BMI as an effort to seize new opportunities made available by the advent of digital technologies. See Foss and Saebi (2017) for details.

2.2.4. Measuring innovation*

Significant efforts have been made in recent decades to measure innovation but it remains a challenge. Innovations that are new to the world are few and far between and one observes more innovations that are new to the market and, of course, even more of those that are new to the firm. Since diffusion is critical for economic impact, it is important to capture innovations that are new to the firm. But measuring innovation at these three levels—products, process and practices—throws up a variety of challenges. While a patent is an *output* measure, R&D expenditure is a standard *input* measure for innovation effort. Figuring out 'novelty' at various levels is the first challenge. Patents are the most common measure of innovation output. However, technically, while patents are an *output* measure of research/technology efforts, they measure *invention* and not *innovation*, as a very small proportion of patents granted are actually used or commercialized. So, if a patent is commercialized, it is clearly 'new' to the world or the market. But, patent quality is not uniform across sectors and nations.

Just like patents are not appropriate to fully capture innovation outcomes, R&D expenditure is also inadequate as a measure of innovation input. Research efforts (R&D) may not result in innovation and consequently R&D is an important but not always essential source of innovation. Moreover, one often does not know what estimates of R&D expenditure actually capture. Reported R&D expenditure can go into salaries, hardware, software, blue sky research,

---

* A lot of literature is available on the measurement issues. The Oslo Manual is the reference point for most of this work. See, for example, OECD (2011) and Gault (2016). This section draws on this literature.

closer to the market research, adaptive research, prototype development and so on. In a worst-case scenario, these estimates are subject to accounting jugglery, especially in situations where firms can claim subsidies or tax credits for such expenditures (Guceri 2016). Given all these issues, innovation is often captured through surveys and is typically *self-reported* by enterprises.

While no detailed time series data is available for innovations by Indian firms, estimates from a large World Bank survey provide useful insights (Table 2.1):

i. As many as 45 per cent of surveyed firms reported *product* innovations, and of these firms, 68 per cent claimed that these innovations were *new* to *their* market. A firm's market can be local, national or international. From a sub-sample of firms, one could ascertain that the bulk of these innovations were *new* to the local market and very few (8 per cent) were reported to be new to the global market;

ii. The proportion of firms reporting *process* innovations was somewhat higher at 48 per cent. As in the case of product innovations, in the majority of cases of process innovations, the novelty was restricted to the local market;

iii. Many firms also reported a variety of *practice* innovations that took the form of *new or significantly improved* (a) logistics, delivery or distribution systems (45 per cent); (b) support systems like maintenance, purchases, etc. (47 per cent); and (c) organizational innovations or management practices (44 per cent); and

iv. In addition, about 47 per cent reported innovations in marketing methods, which may also be considered *practice* innovations but not necessarily in the 'technology innovation' domain.

Table 2.1
**Measures of Innovation-Related Outputs in Indian
Manufacturing and Services Sector, 2013**

Type of Innovation Output	*Percentage of Reporting Firms*
Product innovations (new/significantly improved products)	44.9
a. New to the firm's market	68.2
b. New to the local market*	56.0
c. New to the national market*	19.8
d. New to the international market*	8.4
Process innovation (new/significantly improved manufacturing method)	47.9
a. New to the local market*	49.7
b. New to the national market*	15.7
c. New to the international market*	4.4
Practice innovations–I (new/significantly improved logistics, delivery, distribution methods)	44.9
Practice innovations–II (new/significantly improved supporting activities for processes, e.g., maintenance systems or operations for purchases/accounting/computing)	47.4
Organizational innovations (new/significantly improved organizational structures or management practices)	44.2
Marketing innovations (new/significantly improved marketing methods)	46.9

Source: Computed from World Bank Enterprise Survey—India (Manufacturing Module and Innovation Module).

Sample size=9281.* These questions were asked to a sub-sample of firms and the number of responding firms varied between 2000 and 2500.

As mentioned, many of the practice innovations in Table 2.1 can be seen as BMIs. Moreover, apart from the novelty to the defined market—local, national, international—the nature of innovations in products and processes can vary a great deal in terms of incrementalism, quality features, functions and so on. And these characteristics can vary even for the most important innovations of firms (Table 2.2). New innovations can bring in novel product features, use new inputs, improve quality, reduce costs, build on technology (designs, methods) already available with the firm or get new ones. Innovations based on previous knowledge built in a cumulative manner are an important part of the innovation process—both for products and processes. Besides, process innovations can be critical for product innovations and therefore, both for policy and strategic reasons, the links between product and process innovations need to be understood.

Table 2.2

Some Features of Product and Process Innovations in Indian Manufacturing and Services, 2013

Features of Firm's Most Important New Product or Service	*Percentage of Firms Reporting These Features*
Completely new functions	64.2
Less costly to produce/offer	30.4
Better quality	77.1
Uses different inputs	59.8
Based on technology or industrial design not already used in the establishment	42.0
Features of Firm's Most Important New Process	
Automates manual processes (partially or fully)	61.7
Adapts a technology or method previously used by the establishment	32.7
Introduces a new technology or method	69.9
Uses a more efficient technology or method already used by the establishment	53.8
Most important innovative process associated with an innovative product	70.5

Source: Computed from World Bank Enterprise Survey—India (Innovation Module).

Sample size=2275 (for product innovations) and 2403 (for process innovations).

2.2.5 Protecting Innovation

For a society to benefit from innovations, incentives are required for innovators to pursue innovation-related activities.

A key to this incentive system is that the innovating entity is able to appropriate a significant part of the benefits flowing out of its innovation. For such appropriability, protection of innovation from imitation is critical. Organizations protect their innovation (and consequently the competitive advantage they might derive from it) through various mechanisms. IP protection is one of the most important mechanisms but not the only one. Various types of IP protection are available for different types of innovations and include patents, copyrights, trademarks, trade secrets, etc. (see Box 2.2). It is important to appreciate the fact that these methods of protection may be used in combination, that is, various aspects of a particular product or process may be the subject of different types of intellectual property. For example, a new product can be protected through a patent and if this innovative product is produced by a new process, that can also be patent-protected. But an organization, for strategic reasons, may not disclose all the information about the new product and the process in the patent application and decide to protect it as a trade secret.

Box 2.2

Types of Intellectual Property (IP)

Different types of IP protection are provided by granting exclusive rights to exclude others from exploiting, at least for a limited time, a body of knowledge, an expression, a sign or symbol. IP does not provide the rights to use it without fulfilling other conditions of use/commercialization (e.g., FDA approval) being met.

'Utility' Patent
Patents cover inventions including products, processes and compositions of matter. For patentability, the invention must be *novel, useful,* and *non-obvious to someone skilled in the art.* Usefulness requires that invention has a practical application; it does not refer to any standard of economic value. The emphasis is on *operative technology,* as opposed to *abstract thoughts* or *creative expression.* Typically, protection (patent term) is for twenty years *from the date of filing an application* with the patent office. *Patents, like all intellectual property, are territorial.* That is, the protection offered by a patent applies only within the country which granted the patent. An invention must be described completely in the application in order to fulfil a *disclosure* requirement.

Utility Model or Petty Patent
A utility model (or petty patent) is of a shorter duration than patents—typically four to seven years. The *inventive step* required is minimal and its purpose is to protect minor improvements or adaptive innovations.

Design Patent/Industrial Design

This form of protection applies to an invention's *shape* or *form*, provided these do not serve solely to obtain a technical result. The requirement for protection is *novelty* or *originality* of appearance.

Copyright

A copyright protects the *expression of an idea* rather than the *idea* itself. Protection is for *original* works *fixed in a tangible medium of expression.* These include books, music, motion pictures, computer programs, databases, sculpture, poetry, etc. Exclusive rights are provided for reproduction, distribution, publication, public display or performance and for preparing adaptations or 'derivative works' based on the work, such as movies based on a novel.

Trademarks

Trademarks cover commercial symbols, names or logos, used to identify products, services or their producers. Related *trade dress* protection covers distinctive packaging or labelling, or 'look and feel' of the product. In order to qualify, a trade symbol or dress must be *distinctive, non-functional* and it must have *priority of use* or *priority of registration* with a national trademark office. Trademark protection continues as long as it is used in commerce except when it becomes, through use, a generic term for a product.

Trade Secrets

At its broadest, a trade secret is virtually *any information or expression* in any form which confers upon its owners a *competitive advantage*, i.e., has economic value, and which is *reasonably guarded*. Inventions as well as non-technological information such as financial data or customer lists would qualify. Trade secrets are not defined by law and are not subject to formal requirements as are other forms of intellectual property. They may be protected by physical measures of secrecy as well as by restrictive clauses in contracts with employees, customers, suppliers or other parties to whom the secret is revealed.

Plant Breeder's Rights (PBR) or Plant Variety Rights
For PBR protection, a new plant variety should be *stable* (reproduces true to form over repeated propagations), *homogeneous* (important characteristics are *uniform* across a single planting) and *distinct* (from other varieties). Protection is typically for at least twenty years, or longer for trees and vines.

Layout or Integrated Circuits (IC) Design
Layout design of an IC (semi-conductor)—the *original* and *inherently distinctive* three-dimensional layout or topography of the elements and interconnections of the chip—is covered and protected for ten years.

A major misperception is that IP law (or IP policy) in and of itself constitutes an effective system of protection. In reality, several other components—including a legal, political and economic system—must be in place if intellectual property is to be reliably and appropriately protected. Chief among these is the requirement for a speedy, fair and impartial court system for the resolution of disputes. The court need not be technically competent—technical expertise can be obtained from consultants—but it must be perceived as being even-handed or the system will not function properly. Moreover, studies have shown that IP protection is not necessarily seen as the most effective mechanism of protecting innovation. Other strategies including good manufacturing systems and practices that reduce lead times and improve service quality may also be quite effective (Table 2.3). Insofar as marketing innovations and practice innovations mentioned in Table 2.1 can result in improved times to market or better response times for customer needs, these can be a source of higher appropriability for a firm's innovation.

Table 2.3
The Effectiveness of Mechanisms for Protecting Innovation
Percentage of Innovations for which Different Mechanisms were
Considered Effective

Product Innovations

Industry	*Secrecy (per cent)*	*Patents (per cent)*	*Lead Time (per cent)*	*Sales/ Service* (per cent)*	*Manufacturing* (per cent)*
Food	59	18	53	40	51
Chemicals	53	37	49	45	41
Drugs	54	50	50	33	49
Computers	44	41	61	35	42
Electronic components	34	21	46	50	51
Telecom equipment	47	26	66	42	41
Medical equipment	51	55	58	52	49
All industries	51	35	53	43	46

Process Innovations

Food	56	16	42	30	47
Chemicals	54	20	27	28	42
Drugs	68	36	36	25	44
Computers	43	30	40	24	36
Electronic components	47	15	43	42	56
Telecom equipment	35	15	43	34	42
Medical equipment	49	34	45	32	50
All Industries	51	23	38	31	43

* Shows the percentage of companies that reported that complementary capabilities in sales and service, and in manufacturing, were effective in protecting their innovations.

Source: W.M. Cohen, R.R. Nelson and J.P. Walsh, 'Protecting Their Intellectual Assets: Appropriability Conditions and Why US Manufacturing Firms Patent (or Not)', NBER Working Paper No. W7552 (February 2000).

The important role of lead times apart from access to appropriate manufacturing, marketing, distribution and after-sales service-related assets in ensuring effective protection of an enterprise's innovation is consistent with the key insights of Teece's (1986) on profiting from innovation. His work highlighted the role of appropriability regimes, which are dependent on the stringency of IPRs, and access to specialized complementary assets that are critical for commercializing innovation. Marketing, manufacturing and service-related assets apart from complementary technologies are considered by him to be relevant for profiting from innovation. Building on Teece's core argument, one can postulate that effective use of the linkages between the knowledge embedded in the 3 Ps—products, process and practices—can help enhance the appropriability of an innovation. Process and practice innovations would capture most of the innovations in manufacturing, servicing, etc. Besides, the tacit nature of the knowledge embedded in the 3 Ps would add to its appropriability as many of these linkages are likely to be specific to the circumstances of the firm and cumulatively developed. The 3 Ps framework, therefore, provides strategic insights on how to effectively protect an innovation. Leveraging the linkages between the 3 Ps and the inimitable knowledge that is embedded in the innovation may not only help make imitation by competitors difficult but also give the firm advantage in terms of lead times and access to critical complementary assets needed to take the innovation to the market.

The efficacy of different mechanisms to protect innovation varies not with the nature of innovation—product or process—but also with the type of industry (Table 2.3). For example, patents may be more useful for pharmaceutical drugs and medical equipment industries than for others. What is important to note is that for both *product* and *process* innovations, apart from IP protection provided by patents and trade secrets, capabilities of the enterprise that provide it an edge over its competitors in terms of lead times, manufacturing and service provision, are critical for efficacious protection of its innovations. Better lead times, manufacturing and service quality are often driven by good organizational and other *practices*. Thus, an appropriate mix of product-process-practice innovations combined with adequate IP protection is critical for appropriating the benefits of innovations by an enterprise.

Moreover, firm characteristics may affect the use of various mechanisms to protect innovations. Larger firms tend to use formal mechanisms like patenting for protecting their innovations (Gonzalez-Alvarez and Nieto-Antolin 2007). Indeed, it has been argued that for many innovative small firms, the main issue is not whether to use IPRs or not, but whether to protect their innovations from imitation at all. Most do not use formal IP mechanisms like patents, copyrights, industrial designs, trademarks, etc.; informal modes like secrecy and lead time along with non-protection modes dominate (Thomä and Bizer 2013). Leiponen and Byma (2009) also suggest that within small firms, typically R&D-intensive and science-based small firms were more likely to identify patents as the most important method of appropriating innovation returns. For most small firms, the strategic choice is between secrecy and speed to market.

As one would expect, the nature of knowledge embedded in the innovation would also influence the protection mechanisms. There is some evidence to suggest that firms that predominantly rely on *explicit* knowledge choose patents as a protection mechanism, while those companies that mostly use tacit knowledge tend to opt for industrial secrets (Gonzalez-Alvarez and Nieto-Antolin 2007).

Extant IP regimes can also result in other strategies for protecting innovations. Zhao (2006) shows that technologies developed in countries with weak IPR protection are used more internally, suggesting that firms may use internal organizations to substitute for inadequate external institutions. By doing so, they are able to take advantage of the arbitrage opportunities presented by the institutional gaps across countries. Gonzalez-Alvarez and Nieto-Antolin (2007) show that the method most commonly used by Spanish manufacturing companies is continuous innovation, which enables them to keep a position of technological leadership with respect to their rivals. Such a strategy requires the *practice* of using high-commitment human resources.

Broadly, different protection mechanisms (formal and informal) should not be seen as mutually exclusive as firms combine several methods to form their appropriation strategies. The composition of mechanisms used may vary by industry features, nature of knowledge and firm characteristics. For example, unlike large firms, small firms may not use formal IPRs like patents and if they do, they combine it with informal protection mechanisms.

That the use of IP is not a very widespread mechanism in India to protect innovation is also evident from Table 2.4 which reports the proportion of firms who use various means of IP protection. As expected, the proportion of firms reporting the use of IP is much smaller than the proportion reporting product, process and other innovations (see Table 2.1). Only 16 per cent have applied for a patent for product-related innovations, while this proportion is even lower (12 per cent) for process innovations. The proportion of firms reporting product and process innovations was 45 and 48 per cent respectively. Apparently, many of the innovations are either not patentable as they are only novel locally or firms strategically do not wish to patent as they wish to keep it a secret and use other mechanisms for appropriability. Use of trademark and industrial design is somewhat higher at around 18 per cent each. Use of copyrights is also not very prevalent (14 per cent) presumably because it can only relate to IT and marketing-related innovations.

Overall, therefore, while IP protection is an important policy instrument that helps firms protect and appropriate the benefits flowing out of their innovations, they are often not as effective as other mechanisms. Access to complementary assets to commercialize innovations, for example, is critical for appropriating their value. Insofar as other mechanisms can be impacted by some policy instruments (e.g., through access to finance), the role of policy in enhancing the protection of innovative products and processes of enterprises is not restricted to IP policies alone.

Table 2.4
Use of Intellectual Property by Indian Manufacturing Firms, 2013

Intellectual Property	Percentage of Firms Reporting Use
Applied for a patent concerning a product innovation	15.6
Applied for a patent concerning a process innovation	12.1
Applied for a utility model patent	14.0
Registered an industrial design	17.8
Applied for a trademark	18.2
Applied for a copyright	13.6

Source: Computed from World Bank Enterprise Survey—India (Innovation Module). Sample size=3492.

2.3. Innovation-Related Activities

As discussed, innovation entails a change in the knowledge embodied in the products (services) produced and sold by an enterprise or in the processes/practices employed in this endeavour. Given this perspective, any activity that can generate and/or transfer knowledge about any of the 3 Ps can be viewed as innovation-related as this knowledge can be utilized to develop an innovation. A firm has two broad (but not mutually exclusive) choices regarding the acquisition of knowledge: it can internalize the innovation process by pursuing specific activities to develop knowledge or use existing markets to purchase technology. These choices entail a variety of activities that the firm might undertake. The decision whether to develop technology in-house or purchase it is influenced by benefit-cost comparisons which have to take into account several factors including technology spillovers—the potential and costs of imitation. Consequently, three alternative sources of technical knowledge can be distinguished:

a. Knowledge generated by the firm on its own;
b. Knowledge purchased by the firm; and
c. Spillovers (imitation potential) created by knowledge generation of other entities.

Purchased knowledge can be *disembodied* in the form of technology licences or *embodied* in inputs (including new vintages of capital) the firm purchases. Besides, licences and inputs can either be acquired domestically (within the country) or from foreign sources. In the same vein, technology spillovers can be created from knowledge generation of domestic entities as well as from knowledge generated by foreign firms.

Within these broad sources of knowledge, a wide variety of technology information can be distinguished. The relationship between sources of knowledge can get quite complex if we consider all these sources and their various manifestations. It has been rightly argued that it is important to understand different processes that firms use to manage these sources. In fact, good management of the outside sources of technology information is found to be associated with high innovative performance (Gomes, Kruglianskas and Scherer 2011). If one wants to incorporate this process in our conceptualization of innovation, it can be encompassed in *practices* that companies adopt to manage technology-related information.

Knowledge generated within the firm is usually assumed to be closely related to the firm's R&D efforts. However, R&D is not a homogenous activity. The most common categories of spending are: basic research, applied research and development. Although it is common practice to associate R&D with innovative or technological activity, many other activities in the firm contribute to this process. Mansfield et al. (1971) have shown for some US manufacturing corporations that R&D costs do not account for the major share of expenses or time involved in innovation as the process encompasses research, manufacturing and marketing. Besides, many of the investments in innovation capability-building activities are not captured by R&D expenditures.

Still, undertaking R&D in an enterprise is an obvious activity that contributes to knowledge creation, but incurring R&D expenditures does not necessarily mean that innovation outcomes in the form of new products, processes or practices would emerge. As mentioned, purchase of technology in embodied (e.g., machines, raw materials and components) or disembodied (e.g., through licensing of technology) forms can also generate or provide access to new knowledge relevant for innovation. Hiring of skilled persons, training of the workforce, collaborative arrangements with external entities, linkages with customers and suppliers, unbundling purchased and other available technology can all result in such knowledge flows.

Often, formal and informal linkages with various stakeholders provide crucial knowledge, including feedback for continuous improvements and even significant developments in innovations. There is ample evidence to suggest that a large proportion of ideas for adaptive and innovative changes come from the users of the firms' products.* Despite the fact that there are multiple sources of innovation for a firm, R&D helps organizations to appropriately search for new knowledge and improve their capacities to absorb knowledge from external sources.

As enterprises develop and mature, they tend to broaden and deepen their initial base of innovation capabilities. The cumulative and evolutionary nature of the learning process generally means that firms develop along particular trajectories determined by their initial positions, entrepreneurial strategy and external stimuli. Firms that are able to put in place practices (organizational routines and systems) that facilitate *learning to learn*, i.e., allow them to absorb external knowledge, build on it and create new capabilities, tend to be more innovative and competitive. Insofar as routine business activities may also entail such knowledge flows and learning opportunities, capturing and utilizing them effectively is crucial and the relevant organizational practices are critical for this process as well.

Overall, therefore, a large variety of activities can be categorized as 'innovation-related activities' as they can be

---

* See, for example, von Hippel (2009), Bogers and West (2012), Nga (2020).

a source of new knowledge that is potentially relevant for innovation. Understanding certain aspects of these activities and the nature of knowledge flows they can entail are crucial to appreciating policy-innovation linkages and indeed innovation or technology management strategies.

2.3.1 Measuring Innovation-Related Activities

While a patent is an *output* measure, R&D expenditure is a standard *input* measure for innovation effort. But, as mentioned earlier, just like patents, R&D expenditure is also inadequate as a measure of innovation-related inputs.

It is noteworthy that while the proportion of firms reporting innovations is reasonably high (about 45 and 48 per cent respectively for product and process innovations, Table 2.1), a much smaller proportion (33 per cent) reported that they undertake formal R&D. However, a question about R&D without using the term 'formal' elicits a different response from a sub-sample of firms and the estimate of firms reporting R&D shoots up to about 46 per cent, which is closer to the proportion of firms reporting innovation (Table 2.4). Interestingly, about 48 per cent of the surveyed firms reported that they provide time to their employees to pursue innovation-related activities and between 42 and 47 per cent firms report provision of training to their workers, at times focusing on innovation-related activities.* Interestingly, more than 63 per cent firms reported that they purchased new equipment, machinery or software for innovation, implying that purchase of new embodied knowledge *directly* helped

---

* Usually one would expect a larger proportion of firms reporting general training for their workers than those reporting innovation-related training. But the estimates presented in Table 2.4 are just the opposite. It may be due to reporting errors or due to the fact that the two sample sets are different.

them in developing their product and process innovations. About 10 per cent firms reported that they are *currently using* foreign licensed technology while about 6 per cent had *purchased or licensed* disembodied technology in the last couple of years. Thus, while one may be a bit sceptical about the veracity of the estimates capturing innovative inputs, it is clear that many innovations are either based on informal R&D within the firm, training and/or embodied and disembodied purchased knowledge.

Technology licensing expenditures and purchase of new equipment and raw material are used as proxies for buying *disembodied* and *embodied* innovation (technology), respectively. But here again, degrees of innovation are difficult to capture as the degree of embodiment and vintage of knowledge are not easily measurable. Except in cases where detailed information is available while doing case studies, etc., these expenditures on buying of knowledge often do not capture how much of the knowledge was actually captured.

Table 2.5
Some Measures of Innovation-Related Inputs in Indian Manufacturing and Services, 2013

Type of Innovation Input	Percentage of Reporting Firms
Undertake formal R&D (in-house or contracted with other firms)	33.0
a. Undertake internal R&D*#	46.2
b. Undertake external R&D*#	9.3
Provide time to employees to pursue innovation-related activities	47.5
Provide training to full-time employees	42.5
Provide formal training for innovation-related activities*	47.0
Purchased new equipment, machinery and software for innovation*	63.3
Use foreign licensed technology	10.1
Purchased or licensed any patented or non-patented inventions/knowledge*	6.3

Source: Computed from World Bank Enterprise Survey—India (Manufacturing Module). Sample size=9281.

* These questions were asked to a sub-sample of firms. Sample size=3492.

The question asked in the innovation module was somewhat different as it did not mention the term 'formal'. Consequently, the estimates of the three types of R&D are not strictly comparable.

Apart from undertaking informal and formal R&D and purchasing embodied and disembodied knowledge, firms source innovative ideas from a variety of entities. This informal learning takes various forms including imitation of knowledge that is revealed in the products and processes that are available in the market or through hiring of workers who bring in new knowledge to the enterprise based on their experience elsewhere. Such knowledge flows are referred to as *knowledge spillovers* as enterprises do not directly pay for this knowledge but acquire it through *contagion*—by observing or coming in touch with products, processes and practices of others. Such knowledge flows, however, are not automatic but require efforts at the enterprise level. It is quite well known that for enterprises to benefit from such knowledge spillovers to develop innovations or for competitive advantage through other means, they have to make some efforts on their own. These efforts can take the form of R&D (formal/informal) or technology purchase. Else, they are not able to absorb and effectively use the knowledge available through spillovers (Cohen and Levinthal 1989; 1990). Thus, exploitation of imitation potential by enterprises for innovation often requires investments in making or buying knowledge. Therefore, make-buy-imitate choices for acquiring knowledge are often complementary and not substitutes of each other.

Getting ideas from others (or knowledge spillovers) for innovation by an enterprise is quite common. This is evident from the estimates presented in Table 2.6. More than 50 per cent firms (54–57 per cent, in fact) report that their product or process innovations were either reproductions or adaptations of existing products and processes. This is consistent with the fact that many innovations are only new to the local market and few of them seem to require patent protection. Such innovations may not be triggered by more stringent IP regimes but can be a mechanism to deal with competitive pressures. In fact, prima facie, imitative or adaptive innovations may be curtailed or discouraged by strong IPRs, thereby reducing diffusion of knowledge through such innovations. The innovation vs diffusion dilemma which an IP regime needs to resolve therefore is a real one and we will get back to it in subsequent chapters.

Table 2.6
Developing Product and Process Innovations in Indian Manufacturing and Services: Ideas, Collaborations and Linkages, 2013

Characteristics	*Percentage of Firms Reporting These Features*	
	For Product Innovations	*For Process Innovations*
Source of idea for the most important innovative product/ service/process		
a. Adaptation of an existing product/service sold by another firm	39.6	38.0
b. Reproduction of a product/ service sold by another firm	14.3	19.4
c. Idea originated within the establishment	46.1	42.5
Innovative product/service/ process developed		
a. Entirely by the establishment	83.8	72.0
b. Entirely by another firm	7.5	12.3
c. In collaboration with others	8.6	15.7
i. With a domestic firm	66.2	73.5
ii. With a foreign firm	9.6	10.1
iii. With a domestic research institution	8.6	6.1
iv. With a foreign research institution	3.0	2.4
v. With a private consulting firm	16.7	12.4
vi. With the government	4.6	4.2

Employees specifically hired for developing an innovative product/process	44.6	46.1
Increase in the number of skilled employees as a consequence of new product/process	38.8	40.7

Source: Computed from World Bank Enterprise Survey—India (Innovation Module).

Sample size=2275 (product innovations); Range 2403–2958 (process innovations).

Collaboration with external entities is another activity that an enterprise can undertake to develop innovations. Such collaboration does not seem to be very common in India; only about 9 per cent innovating firms reported such collaborations for product innovations while the percentage was somewhat higher for process innovations at 16 per cent. Interestingly, a significant proportion of firms get their innovations entirely from an external entity; about 8 per cent for product innovations and 12 per cent for process innovations. Among the entities with whom Indian enterprises collaborate to develop innovations, other domestic enterprises are prominent, although foreign firms, consulting organizations, research institutions (both domestic and foreign) and the government are also used as collaborators (Table 2.6).

Finally, the link between human capital and innovation at the enterprise level needs to be highlighted. Skills available to the enterprise are a critical part of its innovation capabilities. Available skills contribute to the innovation process and introduction of an innovation may require a new set of skills at the enterprise. Data suggests that enterprises hire specific types of employees to *develop* innovation (Table 2.6). Hiring such employees can also facilitate learning through absorption and utilization of knowledge spillovers which, in turn, can result in innovations at the enterprise level. If these employees are from competing firms, it can also result in knowledge spillovers as these workers bring in knowledge from firms where they were employed earlier. Similarly, firms may need to hire workers with new skills as they introduce the innovation because employees with older skills may not be adequate for the new requirements. Thus, both for the creation of new knowledge and for introducing it in the market or using it internally, firms require appropriately skilled workers. Non-availability of new skills can adversely affect introduction as well as diffusion of innovation. Skill gaps, therefore, can be a significant deterrent to the exploitation of innovation potential by an economic entity.

The importance of knowledge spillovers is also evident from firm responses on the sources of ideas and information for their innovation activity (Table 2.7). Interestingly, less than 13 per cent firms report their in-house R&D as the most important source of such ideas or information. Customers are the most important source of ideas (24 per cent) followed by suppliers (16 per cent) and the products and services available in the market (15 per cent), which may be getting reverse-engineered. Customers in more demanding markets (e.g., export markets) can be more important as a source of ideas as compared to those located in less demanding ones. Among other sources, business conferences, other firms and consulting firms are somewhat important. The Internet, recent hires from other firms, government programmes and universities/research institutes do get a mention as the *most important* source of ideas but their role seems to be limited. Since ideas for innovation come from a large variety of sources, activities that manage such information flows are critical for innovativeness for an enterprise. The ability, therefore, of an enterprise to leverage such knowledge spillovers may depend on the *practices* it uses to manage and utilize these knowledge flows. Such practices, as mentioned, while defining innovation, are a very important part of the innovation capability of an enterprise.

Table 2.7
Sources of Information and Ideas for Innovation Activity, 2013

Most Important Source of Information or Ideas	Percentage of Firms
In-house R&D and personnel	12.6
Recent hires from other firms	1.2
Knowledge from parent or another firm	5.6
Suppliers	15.8
Consultancy firms	5.5
Business associations and conferences/exhibitions	7.9
Professional journals and trade publications	3.9
Products or services available in the market	14.9
Government ministries or programmes	1.9
Universities and research institutes	0.8
Internet	3.3
Customer feedback	24.3
Do not know	2.5

Source: Computed from World Bank Enterprise Survey—India (Innovation Module), 2014.
Sample size=3492.

2.3.2. Sources of Knowledge for Innovations: China vs India

It is evident that innovation today is located in a network of relations a firm has with a variety of entities; a set of complementary resources are derived from these linkages with myriad mechanisms for information and knowledge flows. A survey of firms in Pune, India and Beijing, China brought out some interesting differences in these linkages, indicating differences in the innovation ecosystems of the two city clusters as well as the two sectors—automotive and software. Table 2.8 summarizes the sources of knowledge for *product* and *process* innovations in auto-component and IT firms. As compared to Pune firms, a significantly larger proportion of firms in Beijing reported employees, suppliers, universities and the government as sources of knowledge for innovations. There are also differences across the two sectors. For example, while suppliers and clients seem to be more important for the automotive industry, returnees (non-resident), competitors, universities and the government are more important for the software industry.

Table 2.8
Importance of Sources of Technology and Knowledge for Product or Process Innovation in China and India, 2007

Type of Sources	*Percentage of Firms*	
Automotive Sector	*China*	*India*
Existing employees (excluding returnees)	73.3	38.1
Returnees from abroad	15.3	4.0
Suppliers	69.0	44.3
Clients	80.5	90.1
Competitors	59.5	53.9
Consultancy companies	20.0	12.5
Universities	24.7	0.7
Government	34.2	7.8
Other	1.1	0.0
All (Number of firms)	100.0 (196)	100.0 (273)
Software ICT Sector	*China*	*India*
Existing employees (excluding returnees)	81.8	37.1
Returnees from abroad	22.2	30.3
Suppliers	36.9	46.2
Clients	76.3	65.2
Competitors	58.1	36.7
Consultancy companies	15.2	40.7
Universities	32.8	14.9
Government	49.0	14.0
Other	0.5	1.8
All (Number of Firms)	100.0 (198)	100.0 (221)

Source: Chaminade (2010).

Within each industry, the differences between the two countries are quite interesting. In Beijing, suppliers are more important for the auto-component industry while the government is a more important source for the software industry. The role of the government in Beijing in the software industry is mainly through public procurement but also through funding laboratories for spin-off firms that develop software applications for defence and other government needs (Chaminade 2010). For Pune on the other hand, competitors are a significantly more important source for auto-component firms, while returnees, universities and consultants are more important for software firms. Moreover, suppliers as important sources of knowledge for Beijing firms suggests that the nature of these linkages is more innovation-related there than in Pune. The differences in the role of universities in the innovation process raise questions about the emerging role of educational institutions in building the innovation ecosystem in India. In recent years, analyses of issues around higher education have highlighted a variety of inadequacies in the higher education system. We will get back to some of these issues in Chapter 4.

2.3.3. *Some Characteristics of Knowledge Embodied in Innovations*[*]

In order to fully appreciate the discussion on the definition of innovation and innovation-related activities, it is important to understand certain key characteristics of knowledge, as managing knowledge is the centrepiece of both innovation outcomes as well as activities that are critical to the innovation process. Early analytical and policy research on innovation had an implicit underlying assumption that technology consists

---

[*] See a detailed discussion of some of these issues in Evenson and Westphal (1995).

simply of a set of discrete knowledge or techniques, which can be fully described in blueprints or manuals and universally applied. This meant that technology can be easily transferred and investments for assimilating external technology innovations or adapting it for local needs are not required. In effect, assimilation was considered to be automatic and costless. Subsequent research, informed by empirical evidence, showed that innovations are *context specific* and sensitive to the circumstances where they get developed. Therefore, the same specifications for an innovation may not be optimal for all contexts. Moreover, knowledge embedded in an innovation cannot be completely codified and expressed in blueprints, manuals and in physical manifestations of inputs and products. This is because much of the knowledge embedded in the 3 Ps that characterize technology is *tacit*, difficult to *codify* and not easily *transferable*. Some refer to this dichotomy as *implicit* and *explicit* knowledge (Jensen et al. 2007).

The circumstantial specificity also results in learning and knowledge accumulation at the enterprise level which is *cumulative* in nature. *Cumulativeness* of the knowledge built around an innovation makes it more appropriable as others cannot imitate it easily. The cumulative nature of knowledge generation and accumulation may also contribute to the tacitness of such knowledge which is not codifiable and only embedded in organizational practices or in the minds of the employees. In fact, it has been suggested that understanding the complementarities between codified and tacit knowledge may be most critical for learning (Nonaka and Takeuchi 1995).

The recognition that there are *tacit* and *codified* elements of knowledge embodied in an innovation and this knowledge is likely to be *circumstantial or context specific**

---

* In a technical sense, machines may need different types of calibration in different circumstances to achieve specific effects. More broadly, which has been *cumulatively* built implies that transfer of knowledge is not costless and would require significant efforts by the transacting parties, but especially by the receivers of this knowledge, to implement an innovation. *This has implications for innovation strategies of firms which need to be focused on continuous learning, as such learning would determine the efficacy of knowledge transfer. This transfer of knowledge can happen when an enterprise is engaged in a collaboration, purchases technology or undertakes activities to exploit spillover potential through imitation. The enterprise, therefore, has to be cognizant of the tacitness and circumstantial specificity of the knowledge being transferred through all these channels. Circumstances or contexts not only vary across regions with different policy regimes, market environments and resources but also across enterprises which are heterogeneous in a variety of characteristics including their paths of evolution. The limitations of codifying knowledge embodied in an innovation developed internally through the enterprise's own R&D will also inform the practices used by the enterprise to diffuse the relevant knowledge internally. The policy relevance of recognizing these knowledge characteristics will be discussed in some detail in the next chapter.

Suffice it to mention here that the transition from 'learning to assimilate' to 'learning to generate' is critical for policymakers as such a transition provides dynamism to the economy and long-term international competitiveness as well. Both effective knowledge transfers from external partners

---

differences in circumstances can take various forms: wage-rental-interest ratios, quality of inputs, physical and social conditions including social institutions or labour-management practices.
* Many studies in the Indian context during the pre-1991 period showed that a large part of technology efforts by Indian firms was spent to adapt foreign technology, either acquired through licensing, import of machines or through reverse-engineering (Basant 1993).

and building of internal technological capabilities acquired through investments over time are crucial for such a transition. If policymakers or enterprises wish to substitute internal efforts with knowledge transfers from external entities, costs and constraints on such transfers, especially of the tacit elements, need to be recognized. This recognition would mean that policymakers create conditions for adequate technology transfer as well as for building internal technological capabilities by firms. Given the characteristics of knowledge embedded in an innovation, simple transfers will not automatically lead to building of capabilities. In fact, even to understand codified knowledge, enterprises require technological capabilities. Given the fragmentation of technology markets, adequate knowledge flows will also not be automatic in various market-driven technology transfer arrangements. Enterprises will need to proactively make efforts to understand the tacit and codified knowledge embedded in the products, processes and practices that characterize an innovation.

Recognition of knowledge characteristics and the role of internal technological capabilities to unravel the cumulatively built, context-specific tacit as well as codified knowledge also adds a useful dimension to the concept of *technology catch-up* in the development literature. The proposition that *technological followers* benefit from the technology created by *technological leaders* was accepted as an empirical truth for many years. It was suggested that the potential for catch-up growth is proportional to the difference in technological capabilities between a follower and the leaders (*technology gap*), predicting a faster growth of entities (firms, nations) with lower technological capabilities. The underlying mechanism for this growth process was *technology transfer*. Bridging the technology gap was seen as somewhat automatic in this scheme of things as adoption and use of the knowledge transferred from the leader took place, supported by investments in education, physical capital and general management capability. Specific and intensive investments in R&D and related activities to build technological capabilities were not considered essential for the catch-up process. Subsequent research has shown that specific investments in technological capability-building over sustained periods are critical for realizing the technology catch-up potential. Apart from other needs, such a capability is critical for unravelling various dimensions of the knowledge available with the knowledge leader. If the technology gap between the leader and the follower is too large, the ability of the follower to learn and catch up may not be feasible.

2.4. Elements of an Innovation Ecosystem

A large variety of participants contribute to the innovation process—individuals, enterprises (both domestic and foreign), higher education institutions (HEI) including research organizations (both public and private), banks, venture capitalists and other funding organizations, government institutions and so on. The linkages between these entities define the innovation ecosystem.

There are three broad and somewhat overlapping conceptual frameworks that inform the analysis of the linkages between innovation and various institutions in an economy, viz., National Innovation System (NIS), triple helix paradigm and University-Industry Linkages (UILs). The concept of NIS was developed to explain the differences in the innovative performances of industrialized countries. More recently, the framework has been applied to analyse the experiences of developing countries, especially the newly industrializing economies of East Asia. Most conceptualizations of NIS have *at least* three components:

a. *Governments* that shape the external environment for innovative activity through a variety of policy interventions and technology-supporting institutions;
b. *Universities and other higher education and research institutions* that supply scientific, technical and other knowledge as well as skills to create and apply new knowledge which are utilized in governments, research institutions and business enterprises (these entities may also generate innovations); and
c. *Business enterprises* which are engaged in development and/or commercialization of new products and processes.

The vibrancy of an NIS is determined by the nature and extent of interactions between the various elements that constitute the system. Typically, these interactions are undertaken within national borders and encompass technical, commercial, legal, financial as well as social transactions.* Given this, the differences in innovative performance across nations are ascribed to the differences in the way institutions combine or interact to generate, improve and diffuse new technologies (innovations)—products, processes and practices. It has been suggested that the literature in this tradition has focused more on the 'invention system' than on the 'innovation system' and has thus accorded less importance to understanding the complementary economic processes needed to convert invention into innovation and subsequent diffusion (Metcalfe and Ramlogan 2008). This idea is similar to the one mentioned above which argues that *all* stages— invention, innovation, diffusion—need to be addressed while evaluating an innovation ecosystem. Such a focus would also

---

* See Metcalfe and Ramlogan (2008) for a succinct summary of the NIS literature.

require the recognition that knowledge embedded in products, processes and practices is complementary and successful commercialization of innovation may be contingent on good understanding of linkages between the 3 Ps.

The triple helix takes the framework of NIS further and focuses on three key helices (actors or elements)—*state, academia* and *industry*—and their relationships.* It develops a 'spiral model of innovation' that captures multiple reciprocal relationships among industry, university and the government at 'different points in the process of knowledge capitalization' (Etzkowitz 2002:02). In doing this, the framework focuses on the 'internal transformation' of each of the helices and the influence of one helix on the other. The framework emphasizes the criticality of the role of academia to generate impetus for innovation which underlies the role of science for creating opportunities for invention and innovation.

The recent literature on UILs appears at first glance to be both an extension and a combination of these two frameworks. However, there seem to be three dimensions where the UIL studies are somewhat different. First, UILs are typically analysed in the context of geographically-bound clusters, related to the literature on regional or cluster-specific innovation systems. In adopting this focus, they tend to de-emphasize the macro-linkages between the educational system and industry. Second, UIL studies focus on the variety of industry-academia linkages

---

* The triple helix framework has evolved over time. In triple helix I, 'the state encompasses industry and academia and directs relations between them'. In triple helix II, the three actors are in individual and separate domains 'with strong borders dividing them and highly circumscribed relations' among them. Triple helix III generates a 'knowledge infrastructure . . . with each taking the role of the other and with hybrid organizations at the interface' (Etzkowitz and Leydesdorff 2001).

and their measurement. Third, these studies are increasingly exploring the complementary economic processes that are required to facilitate and even push the invention-innovation-diffusion process. This exploration has led to the examination of different policy options that bring universities closer to the market and facilitate commercialization of technology developed at the university either through licensing or creation of start-ups. Thus, the emerging focus of this work is to understand the factors that help 'traditional universities' become 'entrepreneurial universities' (Yusuf 2007).

Broadly, all these three streams of literature encompass the linkages between various institutions and innovation; the UIL framework giving more attention to institutions of higher education. Combining the insights from the three strands of literature, it is obvious that paths to the market of knowledge generated are quite complex, multidimensional and often informal and subtle. The initial idea that knowledge generated from academia can be simply transmitted to industry has been replaced by the understanding that it can only be effectively transferred through significant interaction and the knowledge, at times, may need co-creation. Available evidence from India and other developing countries suggests that labour market linkages between industry and academia remain the most prominent link, with universities contributing relatively little to patenting, licensing and new enterprise creation, except to a limited extent in life sciences.* But labour market linkages are quite critical to make the NIS robust and dynamic as education and training help generalize knowledge and embody it in people (Jensen et al. 2007). This knowledge facilitates

---

* See Basant and Mukhopadhyay (2010) and Yusuf and Nabeshima (2007) which will be discussed in subsequent chapters. Some of these trends seem to be changing a bit and we will discuss them in Chapter 4.

both absorption and generation of knowledge. The ability of institutions to respond positively to industry's production and innovation needs and create a virtuous feedback loop depends on the regulatory structure of these institutions of higher education. In the context of the triple helix framework, the academic institutions need to evolve in a particular way in order to respond to such opportunities and create and nurture linkages with industry that can be mutually rewarding. We shall argue in the next two chapters how this is one of the key policy failures as it is complementary to almost all other innovation policy instruments and has a variety of externalities.

2.5. Innovation Process: Forms of Knowledge, Activities and Actors

Irrespective of the framework that one uses, the core of the innovation ecosystem is considered to be business enterprises (Lundvall 2007). Therefore, most studies in this domain recommend that the focus needs to be on designing appropriate policy instruments for enhancing the innovative activities of business enterprises while the other two key elements—government and academia—play a strong complementary and facilitating role. Broadly, two perspectives on these frameworks seem to have emerged over time.* One focuses on the systemic relationships between R&D efforts in firms, S&T organizations (including universities) and public policy. The importance of creating markets for knowledge (e.g., through IPRs) and finance capital (e.g., through VCs) is emphasized here. The other perspective is broader and its conceptualization of innovation processes includes radical as well as incremental innovations and their adoption. Production process and sales

---

* See Chapters 3 and 4 for more discussion on these two sets of studies.

(interactive learning) are important sources of innovation besides science. The importance of learning is critical to both perspectives.

In the wider literature on innovation systems, these two perspectives on NSI can be mapped to two different innovation modes: (i) the mode focused on scientific and technologically-based innovation (STI); and (ii) the mode based on learning-by-doing, by-using, and by-interacting (DUI) (Jensen et al. 2007). The STI mode highlights the relevance of interactions with scientific and academic institutions producing new codified and explicit knowledge through scientific research. This knowledge is used by firms to produce innovation. The DUI mode in contrast emphasizes practice and interaction (often informal) based innovations. Such innovation is essentially generated by the enterprise's capacity to build formal and informal interactions among various stakeholders within the firm as well as with external entities—suppliers, customers and competitors. Such exchanges are critical for eliciting tacit or implicit knowledge. A range of studies have found that a combination of STI and DUI interaction modes has a stronger impact on innovation output than the two separate individual modes (Parrilli and Heras 2016; Jensen et al. 2007). Since business enterprises play the most important role in the innovation system, their modes and intensity of interaction with other firms and with knowledge infrastructure (both domestic and international) are important for innovation outcomes.

With considerable research in the area, the understanding of the innovation process has improved significantly over the years. Almost until the late 1970s, the conceptualization of *technology* in economics was restricted to a process that converts *inputs* into *outputs*. And *innovation* was seen as *change* in *technology*. Technology, innovation and the innovation process, consequently, were all part of the 'black box' that converted inputs into output. The insights on innovation grew as scholars started to 'open' this black box. The sharp focus on the economics of *learning by doing* (Arrow 1962) was one of the initial and critical steps to open this black box. Simply put, Arrow argued that firms learn by production experience. As a result, technology—the relationship between inputs and outputs—changes as enterprises accumulate production experience and their productivity might increase automatically with this cumulative experience. Implicit in the recognition of the process of *learning by doing* is the idea that enterprises learn with experience and can enhance productivity by adapting the technology or using it more efficaciously as they get to know more about it. And, such learning can lead to adaptive and other innovations.

The insights on the role of *learning* in the innovation process became richer once the dimension of *learning by using* was added (Rosenberg 1982). Users acquire critical and relevant knowledge as they *use* the products. This highlighted the fact that the locus of innovation may be *users* as well, apart from *makers*.* Besides, the feedback from users may be quite critical for makers to develop appropriate products or modify them to suit consumer needs. Consequently, the interaction between *users* and *makers* may be critical for the innovation process.

Apart from the fact that innovation is an outcome of different types of learning, capturing the complexity of the innovation process is also difficult because the innovation process is multidimensional. As mentioned, there are different types of innovation—product, process, practice—and within them, radical (major) and incremental. The incremental

---

* The work of Eric von Hippel in this domain is very well known. See, for example, von Hippel (2009). Also see the earlier discussion in the references mentioned in footnote 14.

innovations typically face less uncertainty but being context specific and often tacit, are difficult to document and imitate.

The *interdisciplinary* nature of innovation is another facet that makes it multidimensional. *Engineers* know about what goes on inside the black box, but they focus on the technical issues or problems. But the science of what is going on within the black box may change with significant scientific developments and it may be difficult for the engineers to comprehend these changes easily. Convergence of technologies, with each technology domain getting inputs from multiple scientific domains, makes the innovation process even more complex as different types of scientists and engineers are now required to develop and implement technologies. But scientists and engineers do not focus on the behavioural processes involved in innovation. *Social scientists* do study behaviour and they also write about it, but typically they do not understand the scientific and technical aspects of the innovation process. Managers need to understand both the technical aspects as well as the incentive structures required for innovation. Policymakers' needs are somewhat similar to those of managers as they develop policy instruments to encourage innovation and at the same time appreciate technological specificities of different sectors.

Another feature of the innovation process—unintentional transmission of knowledge to various entities—makes it hard to measure the benefits of innovation. Many innovation outputs have benefits beyond the intended beneficiaries, i.e., they cause *positive spillovers*. And it is important that these are considered by policymakers. Spontaneous, free, direct benefits of innovation to external parties, for which the innovator is not compensated (a low appropriability situation) may not provide appropriate incentives for undertaking innovation-related activities. An innovation may originate in one industry (say electronics), get manufactured in another (say industrial machinery) and used in yet another industry (say textile or chemicals). The differences in *industry of origin*, *industry of manufacture* and *industry of use* of innovations make the exploration of impact even more difficult. This peculiar feature of innovation may mean that policies focused on one sector may effectively affect all three sectors and the policymaker has to be cognizant of these diverse impacts.

2.5.1. Evolution in the Models of Innovation Process

A simplistic linear model of the innovation process, referred to as a *technology push* model, was introduced in the 1950s and 1960s.* In this conceptualization, basic science research provided technological opportunities that were exploited to design and engineer new products which were manufactured and eventually marketed. In this early version of the STI mode, innovation was essentially seen as an application of science wherein knowledge flowed *sequentially* through different stages starting with basic scientific research and ending in sales. In this conceptualization, different people were involved in different stages which meant that there was no overlap or substantial feedback between them.

During the mid-1960s and 1970s, another variant of linear model, called the *demand pull* model, emerged to understand the innovation process. The process in this version started

---

* Several scholars contributed to the early discussions on technology push and demand pull models of the innovation process. The debate was quite lively during the late 1970s and early 1980s. See, for example, Mowery and Rosenberg (1979), Freeman (1982) and Dosi (1982).

not from basic science but by identification of the market need. Such identification of the need led to development of the product and then to manufacturing and sales. Here, the market is the source of ideas for directing R&D efforts; the role of R&D is merely to react to problems identified in the market. The market need changes with incomes, demographics, relative prices and so on. The demand pull model underplayed the role of fundamental science and overemphasized incremental R&D. In some ways, it also highlighted the possibilities of the innovation process getting locked into a cycle where immediate market needs may drive innovations. But like the technology push model, this version of the linear model also viewed innovation as a sequential process with separate stages that do not interact or overlap.

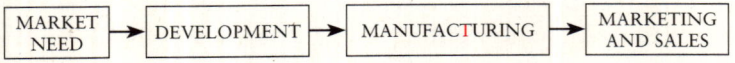

The *linear* models were criticized for their unrealistic assumptions. It was argued that scientists do not have the perfect information about market needs and therefore may not have much to offer in terms of product ideas. The prototypes developed by the design and engineering departments may not be optimal and would require feedback both from manufacturing and consumers to enhance manufacturability and utility. It is quite feasible that the feedback process might bring out the criticality of product design rather than scientific research. Moreover, process and practice innovations that are not relevant in this model may turn out to be crucial as well. Since product adaptation is not considered in this model, issues relating to incremental technological change get ignored.

These models were used for policy prescriptions. Some early studies emphasized the importance of letting market pull shape the emergence of innovations, as demand identifies extant problems that need solutions. Others argued that such a policy focus would mainly result in incremental innovations as existing market needs would receive priority for technology efforts of various stakeholders (Mowery and Rosenberg 1979). It was also pointed out that policies encouraging firms towards market-pull-based innovations would be fundamentally detrimental to national competitiveness as they would take the attention away from basic research, the exploitation of which is critical for long-term economic growth (Mowery and Rosenberg 1979; Freeman 1982). Many, therefore, argued for a combination of the two models (Freeman 1982; Dosi 1982).

Coupling models addressed some of these issues and in some sense combined the technology-push and demand-pull arguments. The core thesis was that innovation is controlled by two distinct sets of forces, namely *technological and scientific progress* and *market forces* which interact with one another in a variety of ways (Kline and Rosenberg 1986). While the former creates new possibilities for developing new products, improving old ones or changing production processes, the latter provides continual changes in commercial opportunities for innovations. These models argued that an innovation model that focuses exclusively on *S&T* might produce excessively complex and costly innovations that are rejected by the market (e.g. the Concorde airplane). And an exclusive focus on the *market* side can lead to the neglect of certain new scientific insights with major potential for practical application.

In their seminal work, Kline and Rosenberg (1986) argued that the essence of an innovation process is reduction in uncertainty through the acquisition of new knowledge and information and aids the learning process. In their perspective, innovation should be seen as a learning process, a perspective

articulated by them in the *Chain-linked Model of Innovation.**
In this model, the process can start either with the perception
of a new market opportunity, an inventive idea based on
new scientific and/or existing technological knowledge or
a combination of the two. Once begun, depending on the
nature of the innovative idea, it is followed by initiation of
research or production of an initial design based on existing
knowledge. Subsequently, based on continual feedback from
R&D, design, manufacturing and marketing, an iterative
process of design and testing ensues, resulting in an innovation
design that is more likely to gain market acceptance.

The *chain-linked model* of innovation (and other
coupling models) has *continuous interaction* as its central
piece. Both *intra-firm* (or inter-departmental) and *inter-actor*
interactions are considered to be important. Interactions of
firms with suppliers, users, competitors, research bodies,
government, universities, scientific institutions, industry
associations, standards institutes, etc., fall in the category
of inter-actor interactions. The processes of innovation and
scientific development interact in various ways at different
stages; there is non-linear feedback on both sides. These
linkages differ across sectors depending on the role of science
in the innovation of these sectors. Overall, all elements
of the innovation ecosystem or the NIS, are considered in
these models. In fact, the *systems of innovation* concept,
which informs the NIS framework, is based on interaction
and learning, and has its roots in the coupling models of
innovation. The critical question is: Can policy facilitate
such interactions?

---

* A lot of subsequent work has been done on this model. We
describe the bare-bones version. For a recent review, see Micaëlli
et al. (2014).

Three major (interrelated) trends have been observed worldwide since the emergence of chain-linked models: (i) increase in competitive pressures associated with the rise of new technologies (e.g., ICT and more recently, machine learning [ML], artificial intelligence [AI], data analytics, etc.); (ii) increase in the convergence of technologies and sectors due to the multidisciplinary nature of new technologies, making the innovation process quite complex as well as interactive; and (iii) profound changes in the worldwide regulatory regime, due to harmonization of trade, investment and intellectual property rules. As a consequence, *speed to market* is increasingly crucial for competitive advantage which requires further integration of innovation stages for better co-ordination through cross-functional teams and maybe, flatter organizations. More specifically, enlightened enterprises are ensuring closer links between product, process and practice (organizational) innovations and moving towards *parallel developments* in the 3 Ps through the use of techniques like design for manufacturability and/or of new technologies mentioned above. The key idea is to achieve greater flexibility and adaptability in products, processes and practices (organizational routines). Participation in innovation-collaboration networks and strategic alliances along with co-development of innovations with users and suppliers are also part of the responses to these three key changes.

2.6. Concluding Observations

In sum, the innovation process has become very complex, enhancing the need for *continuous learning and absorptive capacity* for competitiveness at all levels of the organization. The insights of the coupling models, especially the chain-linked models, which recognized the complexity of interactions across stakeholders and the importance of such interaction for learning, are relevant for both the policymakers and the business enterprises. Recent studies using triple helix, NIS or UIL frameworks, as well as those that focus on STI and DUI modes of innovation, also suggest that chain-linked models capture the empirical reality of the innovation process quite well. The interaction between different entities within the NIS is critical for innovation outcomes. The success of innovation policy is likely to depend on how it is not only able to encourage and facilitate the flow of knowledge across the NIS of an economy but also improve the chances of it being combined and implemented in innovative ways (Dodgson and Gann 2010). The possibility of combining ideas from different sources improves dramatically if organizations have capacities to receive and use the knowledge embedded in these ideas. Therefore, policy also needs to facilitate building of such capacities which can be termed as innovation capabilities. Such capabilities include the ability to develop or search for, select, configure, deploy and learn about knowledge embodied in an innovation. As discussed in this chapter, these capabilities would entail skills, processes and organizational practices that create and use embodied and disembodied knowledge contained in innovations.

In the context of this overview of innovation and innovation activities, the next chapter explores various dimensions of innovation policy.

3

Innovation Policies
Insights from the Literature

3.1. Introduction

An appropriate conceptualization of innovation and innovation policy is critical for developing an understanding of the policy instruments that can potentially affect the nature and extent of innovative activity in an economy. In the last chapter we discussed various nuances of innovation and innovation process. This chapter builds on that discussion and analyses a large body of theoretical and empirical literature to explore the contours of innovation policy. No claim is made for being exhaustive in the coverage of available studies. The key purpose of this chapter is to highlight the complexity of studying the impact of public policies on innovation activity and its outcomes.

The most important role of policy is to establish an environment that stimulates firms and other economic entities to engage in innovation efforts and build capabilities, and introduce or adopt innovations to improve productivity. A key objective of innovation policies is the development of local innovation or technological capabilities. The need for such capabilities is highlighted by the discussion in the last chapter which suggested that not all elements of technology are perfectly tradable, especially the ones that are tacit, cumulatively built and circumstantially specific. The availability of technological information or of innovations somewhere in the world does not mean that these are accessible to everyone; the transaction costs of acquiring this information can be quite high. Besides, the international as well as domestic markets for many elements of technology are not competitive. Therefore, technological capability of the purchaser of innovation greatly influences what is sought and obtained, as well as the price paid and the benefits gained.

In what follows, we discuss the scope of innovation policy and the processes through which various policy instruments are likely to impact innovation.

3.2. The Scope of Innovation Policy

Drawing on the discussion in the last chapter, innovation policy could be seen in two ways. One, the state's *direct* contribution to the NIS through a set of institutions that help generate and diffuse innovations; and two, the set of policies that *directly* relate to science and technology activity of enterprises and related institutions. Broadly, two types of roles for *conventional* innovation policy are recognized:

i. To create and maintain a legal environment conducive to *private-sector investment* in innovative activities. These measures (e.g., IPRs) are expected to enhance the appropriability of innovative activities like R&D; and

ii. To provide financial or non-financial incentives to enhance the level of innovation-related activity. This can take a variety of forms ranging from governmental grants and contracts to tax incentives.

But in a broader sense, any policy instrument that influences an economic agent's decision about *development, adaptation and adoption (through purchase and/or imitation) of an innovation* can be seen as part of the innovation policy. An innovation policy, therefore, needs to address all stages of technological change—*invention, innovation and diffusion*—and all types of innovation. In other words, any policy instrument that impinges on innovation-related decisions of an economic entity is a part of innovation policy (see Box 3.1).

Box 3.1
Narrow and Broad Conceptualization of Innovation Policy

Conventionally, policies that improve appropriability of investments in innovation, provide incentives to undertake innovation activities and create S&T institutions are considered to be part of innovation policy. In this *narrow* conceptualization, policies relating to S&T (including allocation of budgetary resources for S&T), IPRs, tax and other benefits (e.g., subsidy) for R&D and support of S&T institutions would be considered part of innovation policy. All these policies support innovations in various ways. *IPR policies*, for example, affect incentives to innovate as well as the trade-off between making, buying and copying as

IPRs not only create technology markets (purchase/license) through reduction in information asymmetry about innovations and defining of property rights but also define imitation (spillover potential). Diffusion of innovation is affected as market creation and transfer of tacit knowledge is balanced by appropriability through monopoly rights.

A *broader* perspective includes all policy initiatives that can influence innovation-related decisions of economic entities. Several policies can then be seen as *instruments* of innovation policy.

Financial policies: Affect the ability to raise money to do R&D, buy technology and commercialize technology-based products. Financial subsidies for specific types of innovations can help adoption (or impede it if older technologies are supported) and availability of venture capital can support innovation-based start-ups.

Trade policies: Affect the ability to import innovative products, equipment or raw material. Competition from imports may enhance innovation-related investments. Export orientation exposes firms to more advanced innovations (3 Ps) and can facilitate learning through spillovers.

Education policies: Affect availability of skills for innovation development, absorption and adoption. May also affect scientific developments and the role of academia in the innovation process.

FDI policies: Affect type of MNC entry in specific sectors and their willingness to bring new technologies to host countries. This creates potential for local firms to learn from MNCs through linkages and spillovers. Competitive pressures built due to the entry of MNCs may affect innovation-related efforts of domestic firms.

Industrial policies: Affect competition conditions faced by firms through entry/licensing policies which in turn may influence innovative efforts. *Sectoral* or *high-tech* thrusts can create focus on innovation in specific areas. MSME policies may influence linkages between small and large firms and the associated knowledge flows.

Location/cluster policies: Create opportunities for economic entities to network in specific clusters facilitating formal or informal knowledge flows and associated feedback processes and innovation spillovers.

Labour policies: Affect choice of capital or labour-intensive innovations. Incentives for training and ICT adoption may lead to improvements in the 3 Ps.

Policies for UILs: Affect incentives for various stakeholders—institutions, industry and faculty—through IP ownership, patenting support and licensing norms. Creation of incubators in educational institutions and seed funds may support innovation-driven start-ups by faculty and students.

Standards policies: Setting standards reduces regulatory uncertainty and firms can make optimal choices about the technology foci for innovative efforts. The role of standards differs in different stages of the product and technology life cycle and affects the diffusion process.

While a more elaborate discussion on the role of various policy instruments on innovation will be undertaken in a later section, the broad conceptualization of innovation policy as exemplified in Box 3.1 suggests that any policy that changes the nature of competition (e.g., through industrial, trade or competition policies) or modifies access to embodied or disembodied technology can be seen as a part of innovation policy. Moreover, insofar as policies affecting human resource development and UILs influence the ability of a nation to generate and diffuse innovations, these policies are also a component of innovation policy. In the same vein, policies relating to standard setting, incubation of innovation-driven start-ups and financial market mechanisms (e.g., venture capital) can facilitate development and diffusion of innovations. All these have emerged as critical non-conventional instruments of innovation policies.

To briefly recapitulate, a technological innovation characterizes a change in technology. If technology is conceptualized as knowledge embodied in the *products* firms make, the *processes* (e.g., machinery) they employ to make them and the *practices* (or organizational routines) they use to manage the product-process combine, innovation can be seen as a *change* in any of the 3 Ps. Such conceptualization is appropriate for the emerging economies for a variety of reasons. Consequently, instruments that affect firm efforts with respect to any of the 3 Ps need to be included in our broad conceptualization of innovation policy. Before getting into the details of the public policy-innovation interface, it needs to be recognized that different policies do not work in isolation.

3.2.1. Complementarities of Policy Instruments

The last chapter argued that various frameworks of analysing innovation systems and processes emphasize the continuous interactions between various stakeholders. Following that broad idea, studies have argued that the locus of innovation should be seen as a 'network' of organizational relations as firms get into multiple linkages (with other firms, universities, etc.), and acquire technologies and equity in other entities to access a complementary set of resources (see, e.g., Arora and Gamberdella 1990; Rothaermal and Hess 2007). In the same vein, it was suggested in the last chapter that innovation and production strategies of firms are linked, and often product and process innovations are related to each other. Given these complementarities, public policies that provide incentives for one strategy should take into account 'externalities' of such policies for other decisions of the firm (Miravete and Pernias 2006). Therefore, complementarity of different innovation policy instruments may be critical for enhancing the efficacy of these instruments. Broadly, innovation policy instruments need to be coordinated to ensure that firms have access to financial resources for innovation (often an uncertain activity), appropriately skilled manpower, opportunities for collaboration to benefit from 'networks' and adequate information on technology and markets (Strube and Resende 2009).

3.3. Innovation-Public Policy Interface[*]

Public policy influences strategic choices of economic agents regarding knowledge generation and acquisition, which in turn affect innovation outcomes. More specifically, various public policy instruments affect the *supply* and *demand* of knowledge (innovation)[†] and the *competition* and *contagion* conditions that firms face. These impact behaviour of entities that produce, distribute and apply knowledge (see Box 3.1). For example, innovation-related activities of domestic firms may get enhanced when FDI policies are liberalized and foreign firms enter the market. The *contagion* effects or spillover benefits would occur when MNCs demonstrate new technologies, provide technical assistance to their local suppliers and customers, and train workers and managers who may later be employed by local firms. At the same time, *competition*-related pressures exerted by the foreign affiliates may also force local firms to operate more efficiently and introduce new technologies.

Given the perspective that public policy can influence innovation through its impact on *supply* and *demand* of knowledge or through changes in *competition* and *contagion* conditions, virtually all policies can be seen as instruments of innovation policy. As examples in Box 3.1 show, policies relating to trade, industry, FDI, labour, competition and industrial clusters can influence firms' innovation-related decisions. At the same time, specific policies that provide R&D support, undertake public procurement of certain innovative products, finance innovation-driven start-ups,

---

[*] Part of the discussion in this section draws on Basant (2018).
[†] For an excellent summary of technology policies that affect supply and demand for innovation, see Steinmueller (2010).

create standards, help build UILs or influence conditions of technology purchase (e.g., licensing) also affect firms' strategies vis-à-vis innovation.

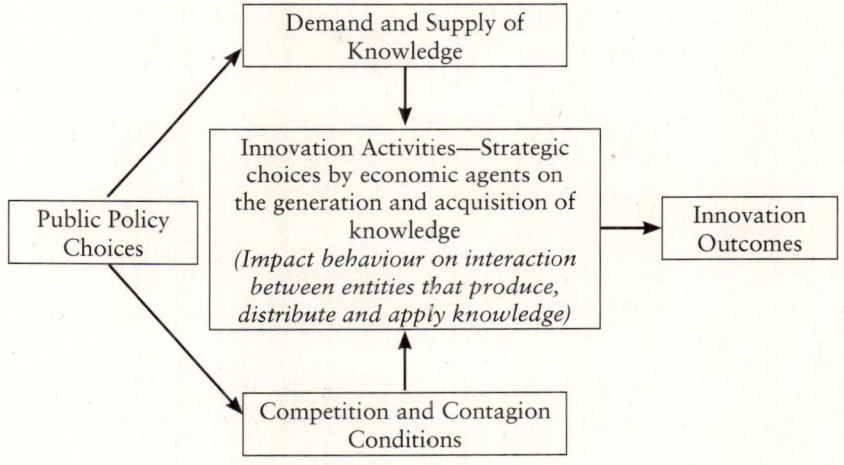

Figure 3.1: Innovation-Public Policy Interface

The interaction between public policy choices and innovation outcomes, however, is a complex phenomenon (Figure 3.1). The challenge is to find appropriate ways of conceptualizing the relationship between innovation and public policy. This would not only require getting inside the 'black boxes' of *innovation* and *public policy* but also that of *firm choices*. The complexity of the relationship to be explored increases as we open different boxes with a set of questions: How can policies influence firm choices about innovation activities? What implications do different policies have for the *supply* and *demand* of knowledge and as a consequence on these choices? How do policies change *competition* and *contagion* conditions and in turn the choices and activities of economic entities? With the definition of innovation and innovation activities provided in the last chapter, we now move on to answering this set of questions.

3.3.1. Supply Side, Demand Side, Competition and Contagion Effects

Policies can simultaneously affect supply-demand conditions for innovation as well as competition-contagion conditions faced by firms. Therefore, at times it is difficult to isolate the effect of policies through these two routes.* But despite such overlaps, it is analytically useful to distinguish between these two mechanisms through which policies can impact innovations.

Driven by strategy, policy or both, innovation-related choices of economic entities affect demand and supply of knowledge in an economy. A major issue vis-à-vis the supply of new knowledge is the quantity and quality of R&D undertaken by economic entities—both in the public and the private sector. The public-good character of new knowledge and its positive externalities result in situations wherein the creators of knowledge are not able to appropriate all the economic benefits of new knowledge and therefore underinvest in R&D. Consequently, weaker competitive conditions may encourage the supply of new knowledge but might discourage its demand. Alternatively, monopolists or oligopolists may have fewer incentives to innovate as they already control a large part of the market. Consequently, absence of competition may result in less innovation. The links between competition and innovation are therefore quite

---

* The rich discussion in Steinmueller (2010) on a variety of policies brings out some of these overlaps quite well.

complex and may depend on a variety of other factors (see discussion below).

In order to correct such market failures in the creation of knowledge, the government uses a variety of policy initiatives in the form of tax credits, protection of IPRs and funding of research (especially at an early stage, commonly identified as basic and applied research). Government grants to directly fund innovation activities or promises to procure new products can enhance the demand for innovation. Policies that provide access to inputs that complement new knowledge can increase demand as well as supply of innovation. These inputs include skilled workforce, modern infrastructure (especially transport and telecommunications), etc.

The make-buy-imitate choices are also affected by competition and contagion conditions, which in turn are influenced by the 'appropriability regime' faced by the enterprise and the prevailing policies. For example, if the IP regime is weak, possibilities of imitation (contagion potential) are high and firm choices may gravitate more towards imitation rather than making or buying. Similarly, with a liberal trade policy, competition from imports would go up and firms may invent or innovate to meet this competition. But in the process of meeting this competition, they would have lower incentives to undertake make/buy activities if the IP regime is weak as they will not be able to reap 'monopoly rents' from such activities, imitation possibilities being high. Thus, a combination of contagion and competition conditions that are affected by prevailing policies would impinge on make-buy-imitate choices of the firms.

Overall, therefore, while the existence of such linkages is well established, the nature of these relationships is quite complex and requires a lot more research. For example, are the links between levels of competition and innovation non-linear? How much competition is 'appropriate' to create an optimal mix of invention, innovation and diffusion? This is a very pertinent policy question, as we have suggested earlier that the role of innovation policy is to appropriately influence all the stages of technological change.

It is well known that the contagion effect (imitation potential) increases with larger pools (supply) of knowledge to learn from, and firms may undertake technological activity to benefit from these contagion effects (Cohen and Levinthal 1989).* In fact, for India as well, some studies have shown that the availability of pools of knowledge or knowledge spillover potential (through R&D and patenting) generated by foreign and domestic firms, increases the R&D efforts of domestic firms (Basant 1993; Basant and Fikkert 1996). Thus, if a more stringent IP regime results in a significantly higher focus on inventive or knowledge-creating activities, the knowledge pool available for imitation would expand. One can, therefore, visualize a situation where a tightening of the IP regime may increase the costs of imitation but result in a much larger pool of knowledge available for imitation. In the presence of a stringent IP regime, technology markets may experience fewer failures with innovative firms willing to transfer technologies as the probability of imitation is lower.

---

* The core argument of Cohen and Levinthal (1989) is that the spillovers associated with imperfect appropriability may actually increase R&D in the industry equilibrium. They argue that there is a positive effect of spillovers on the marginal productivity of the firm's R&D as the firm's own technological effort improves its ability to assimilate the technological developments of others. Therefore, if this effect is sufficiently strong, it can overcome the disincentive of imperfect appropriability, resulting in an aggregate R&D higher than the level it would have reached in the case of perfect appropriability.

In fact, more recent vintages of technology and more tacit knowledge may get transferred. Consequently, the overall impact of a stringent IP regime may be difficult to predict.[*]

These insights imply that 'making' and 'copying' are not fully substitutable; they complement each other to varying degrees. A few research questions that have remained somewhat underexplored in this domain are: How do contagion and competition effects interact with each other? What kind of role does policy play in this process? How do policy choices affect and are affected by the mediating effect of local capabilities? Before we get into a discussion on some of these questions, it is useful to recall that ideas of innovations in enterprises come from a wide variety of sources, reflecting a significant role of knowledge spillovers or imitation potential in the economic environment of the enterprises, which includes its competitors, collaborators, customers and suppliers (see Tables 2.6 and 2.7). Arguably, the role of *contagion* is significant in the innovation process. The role of *competition* is also evident when we explore the reasons why enterprises introduce product and process innovations (Table 3.1). Apart from the nudge created by regulation and standards, all the other reasons can be seen as those which affect the ability of the firm to respond to competition (and consequent reduction in demand) or create competition for others. This gets done by enhancing flexibility, reducing lead times and costs, entering new markets through product innovations, diversifying and expansion.

---

[*] The discussion in the last chapter also suggests that the impact of the contagion and competition conditions on innovation is mediated by innovation and absorptive capacity of the focal firm as well as the distance of the firm from the technology (technology gap) that needs to be learnt from. See Kokko, Tansini and Zejan (1996); Girma, Gong and Gorg (2008) and Kathuria (2010).

Table 3.1
Reasons for Introducing Product and Process Innovations in Indian Manufacturing and Services, 2013

Reasons for Introduction	Percentage of Firms Reporting These Features
Product Innovations	
To replace a product or service offered by the establishment	25.6
To extend the range of products or services offered by the establishment	91.4
To open new markets or increase market share	84.6
To decrease costs of production or service offering	36.5
To offer products or service already offered by competitors	74.0
To comply with regulations or standards	49.8
To deal with a decrease in demand for other products or service	38.8
Process Innovations	
To improve the quality of products or services	79.5
To increase production or services offered	65.7
To increase the flexibility of production or offering service	60.0
To increase the speed of production or offering service	64.3
To improve the speed of delivery to the customer	48.3
To decrease cost of production	45.0
To reduce waste or errors	57.4
To comply with regulations or standards	44.5

Source: Computed from World Bank Enterprise Survey—India (Innovation Module).

Sample size=2275 (Product Innovations) and 2403 (Process Innovations).

In summary, therefore, the role of different policy instruments in innovation is mediated by their impact on supply-demand of knowledge and competition-contagion conditions. Innovative efforts of economic entities are affected by how policy is able to change these two dimensions of their environment. The interplay between policy, enterprise conduct regarding innovation and innovation outcomes becomes even more complex when we recognize the role of enterprise action on their environment.* This recognition would suggest that the arrows in Figure 3.1 connecting firm activities with these two dimensions of the environment should reflect two-way causation. With this broad understanding, we now turn to the discussion of specific public policy instruments and their links with innovation.

3.4. Specific Policies and Their Potential Impact on Innovation

As mentioned above, if we view policies as initiatives that simultaneously change supply and demand of knowledge as well as the competition and contagion conditions, virtually all policy instruments can influence innovation-related activities. Policies that liberalize trade, FDI, industrial policy, infrastructure, etc., typically enhance internal as well as external competition and facilitate knowledge flows. Entry becomes easier, making the markets contestable and

---

* This is akin to enterprise *conduct* affecting the *basic conditions* and the *market structure* it faces; a kind of feedback loop in the modified Structure-Conduct-Performance (SCP) framework of analysing an industry. In the initial versions of this framework, basic conditions and market structure were exogenously determined by policy and other factors and enterprises just responded to them (see Scherer and Ross 1990 for details).

contagion possibilities emerge with more trade and FDI. The impact of economic liberalization on innovation-related investments is difficult to predict as a variety of factors are at work, including the extant degree of market competition and firm resources and size. It may stimulate or reduce innovation investments, depending on how liberalization affects market competition faced by firms. One firm response could be to face competition by investing in new products (to differentiate their products from those of competitors) and more efficient production modes (to reduce production costs and consequently, prices). But if the needed innovation investments are high, innovating firms would require at least temporary market power to appropriate rents and cover their upfront innovation investment costs. Such appropriation may not be possible in highly contestable markets. Aghion et al. (2005; 2006), therefore, propose a U-shaped relationship and argue that in less competitive markets, an increase in competitive pressure may stimulate firm innovation but in more competitive markets an increase in competition may result in fewer firms investing in innovation-related activities.*

Moreover, firm size would also matter for firms' investments in new products and processes if these impose large upfront costs. Access to such resources is likely to be better for large firms and scale permits use of new knowledge at low marginal costs. Cohen and Levin (1989) suggested that in situations where constraints on scale are lower than the pay-off from innovation investments relative to the costs, firms will not invest in innovation. A similar argument can be built for economies of scope as well. If firms are diversified in such a manner that innovation in one segment can be

---

* Some recent studies have contradicted the empirical findings of Aghion et al. (2005). See, for example, Correa (2012).

leveraged in other businesses, firms may be more inclined to undertake innovative activities when faced with competition.

Given this broad context, we discuss briefly the potential impact of specific policies on innovation and explore a variety of industry and firm-level conditions that might mediate these relationships.

3.4.1. Industrial Policy

Apart from other things, *industrial policy* affects entry and exit barriers and hence, the degrees of contestability or competition in the market. Therefore, based on the discussion of the role of competition above, a liberal industrial policy creates a potential of innovation-driven entry and innovation by incumbents to deter entry or meet competition. Similarly, exit conditions like bankruptcy laws may affect experimentation and choice between radical and incremental innovation. *Ceteris paribus*, risk-taking through innovative activity may decline if exit conditions are onerous.

In a very interesting study, Aghion et al. (2014) built a theoretical argument and then tested it with laboratory experiments to analyse the impact of competition on innovation. They found that an increase in competition leads to a significant increase in R&D investments by 'neck and neck' firms, i.e., firms that operate at the same technology level. However, increased competition decreases R&D investments by firms that are lagging behind, in particular if the time horizon is short. It is possible, therefore, to hypothesize that firms may not invest in R&D despite an increase in levels of competition because they face a significant 'technology gap' vis-à-vis the frontier or the competitor. However, invoking the make-buy-imitate choices available to firms, an alternative hypothesis could be that R&D investments do not get undertaken because buying or imitation options are now available.* More empirical work on such processes that explicitly recognize firm heterogeneity in exploring the links between competition and innovation would be very useful.

3.4.2. Foreign Direct Investment Policy

A lot of work has been done on the role of *FDI policy*, innovation and productivity benefits for the host country firms. FDI influences both *competition* and *contagion* conditions as MNC entry adds to competition in the market and at the same time the knowledge that is brought in creates spillover potential. There is some evidence to show that the nature of contagion and competition effects vary with type of FDI (greenfield, brownfield, mergers and acquisitions [M&A] or other); type of MNC ownership (wholly owned subsidiary, joint venture or equity alliance); hierarchy' of activity in which the MNC is involved (R&D facility, contract R&D, manufacturing, marketing and distribution, etc.); technology intensity of the sector (high-tech or low tech); and the nature of technology flows (embodied or disembodied, tacit or codified) (Meyer 2003). Typically, greenfield entry, wherein the MNC not only creates new capacity but also brings in new vintages of knowledge, has advantages that are not available with the other entry modes. More knowledge may get transferred to a wholly owned subsidiary due to lower appropriability concerns, but JVs or equity alliances may provide more opportunities for the domestic partners to learn, even though learning potential may be less given the lower quantum of knowledge flows.

---

* It has been argued by some studies that similar processes were at work during the early phases of economic liberalization in India. See discussion in Chapter 4.

If MNCs participate in low-end activities that are not knowledge-intensive, they may bring in limited knowledge flows to the host country limiting learning potential; manufacturing and R&D activity are better in this respect but MNCs may make special efforts to reduce knowledge transfer especially in R&D. Similarly, the presence of MNCs in high-tech sectors and flows of tacit knowledge would be more useful for the host economy. However, empirical results on the impact of the characteristics of FDI have not always been consistent.*

It has been suggested that the growth of research capacity in nations like China and India in recent years has the potential to change the ways in which MNCs organize their R&D activity (Dodgson and Gann 2010). Usually, MNCs create overseas R&D facilities to adapt their products for local markets, benefit from local research expertise and build global networks of research collaboration. The nature of activity undertaken by the R&D labs in the host country will affect the flows of knowledge and the consequent impact on local innovation capabilities. Typically, an adaptation focus might link the work of the MNC lab to local markets and result in local knowledge flows. If the R&D lab is an important component of the MNC's global R&D efforts, the level and complexity of R&D activity may be high but flows of knowledge within the host economy may be low if the R&D activity is only integrated with the core R&D efforts of the parent company with no local linkages.

---

* Stiebale and Reize (2008), for example, show that after controlling for endogeneity and selection bias, foreign takeovers have a negative impact on the propensity to perform innovative activities and on average R&D expenditures in innovative firms. Takeovers also do not seem to result in significant technology spillovers. Some other studies tend to show opposite results (Johansson and Loof 2005).

In recent years, while some studies on the role of FDI in technology transfer and spillovers suggest that the impact is somewhat unclear (Iwasaki and Tokunaga 2016; Crespo and Fontoura 2007), others argue that FDI inflows increase R&D and innovative activities in host countries (Erdal and Gocer 2015). Still others suggest that the impact of FDI depends on the nature of linkages (backward, horizontal, forward), absorptive capacity, technology gap (referred to above) and institutional characteristics (Gorodnichenko, Svejnar and Terrel 2014). There is also evidence to show that the impact of FDI on technology spillovers varies *with time*; it is initially negative but has a *permanent* positive effect over time (Merlevede, Schoors and Spatareanu 2014). Proximity of domestic firms to MNCs also helps (Wang and Wu 2016). An interesting recent study using data on US MNC affiliates shows that the distribution of FDI in R&D differs from that of general FDI, but increasing value addition by MNCs predicts foreign investment in R&D in the future. In other words, FDI in R&D is an upgrade decision (Wellhausen 2013).

Given the complexity of the ways through which FDI may affect innovation activity in the host country, policymakers need to worry about the nature of MNC involvement that various policy instruments would entail and their impact over time. They also need to keep sight of the fact that most studies reiterate the earlier findings that good absorptive capacities of domestic firms and of the regions where MNCs are located are preconditions for benefits to accrue from positive FDI externalities (Crespo and Fontoura 2007).

3.4.3. Trade Policy

Trade policy choices affect access to embodied innovations through imports and competition through imports-based entry.

The contagion potential is affected by the technology or skill intensity of imports (machines, components, raw material) as that provides the basis for knowledge spillovers-based 'learning'. With trade liberalization, entry through imports is typically not affected by several entry and exit barriers that a domestic enterprise usually faces. Consequently, such an entry can be of a 'hit and run' variety, resulting in high competition effects. Trade policy also modifies 'entry' choices for foreign firms; with very liberal trade policy, MNCs can potentially enter through exports rather than through equity-based entry.

In addition, excessive protection against imports can destroy the incentives for producers to innovate or even search for the most appropriate technology. Exports present an opportunity for firms to learn from demanding foreign buyers and improve products, processes and practices to become more competitive. Thus, import substitution policy reduces trade competition and export promotion (even an undervalued exchange rate) improves both competition and contagion potential because, through export market participation, domestic firms can benefit from learning by exporting along with associated knowledge flows and exposure to competition in foreign markets. One can argue, of course, that import substitution can also create markets for domestic players to build technological capabilities due to a protected market. The East Asian and Chinese experience would suggest that *temporary* import substitution and export promotion with clear and transparent performance targets can potentially build innovation capabilities. In fact, export-led growth in these economies was probably the most important driver of building absorption and innovation capabilities as they competed and learnt from very demanding markets both in terms of price and quality. Initially, export participation facilitated a variety of process and practice innovations but over time, product innovation capabilities also built up.*

Broadly, trade liberalization can seem to be different from industrial liberalization as it offers larger opportunities for technology spillovers (contagion potential) through the mechanism described above and exposes firms to competition both in the domestic and external markets, apart from providing access to the larger global market via exports. Empirical studies have shown that trade liberalization policies are positively related to firms' innovation activities (Amiti and Khandelwal 2013; Bustos 2011; Fernandes and Paunov 2013). There is some evidence to show that export market participation increases investments in R&D and training to develop capabilities for absorbing foreign technologies. It also enhances sourcing and use of advanced technologies and improves the novelty of the firm's innovation (Baldwin and Gu 2004).

There is also some evidence to show that the pre-reform (1991) India trade protectionism adversely affected technological activity among Indian firms and higher rates of protection discouraged firms from keeping abreast of recent technological developments through making or purchasing technology (Basant and Majumder 1997). A recent paper introduces firm heterogeneity based on efficiency levels, in a theoretical model to explore the impact of trade protection on innovation activities (R&D) of firms. The results suggest that trade protection provides incentives to undertake R&D and associated productivity benefits to *less* efficient domestic firms while the *highly* efficient firms see a reduction in R&D and

---

* Several studies have analysed the East Asian and Chinese growth experience. See, for example, Amsden (1989), Lall (1994) and Yusuf and Nabeshima (2010).

productivity with such protection (Song and Vandenbussche 2008). In this model the less efficient firms undertake R&D presumably to catch up. It is not clear whether a large technology gap would constrain catch-up processes and if the outcomes would be different in the absence of domestic competition. While the role of domestic competition is still unclear, a very interesting study by Amiti and Khandelwal (2013) finds that lower tariffs are associated with quality upgrading for products *close to the world quality frontier*, whereas lower tariffs discourage quality upgrading for products *distant from the frontier*. In other words, a large technology gap may discourage innovation efforts in the presence of import-based competition.

3.4.4. Policies for Education and Science and Technology

Education and S&T policies can intensify competition by generating knowledge for innovation-based entry and facilitate contagion by building capabilities to absorb technology and exploit spillover potential. Existence of technological capabilities (e.g., trained S&T personnel) can attract innovation-intensive investment, both domestic and foreign, thereby enhancing competition as well as contagion potential. Such capabilities can also enhance participation in global innovation networks which also facilitates learning (contagion). Insofar as these policies can also create incentives for commercialization of technologies developed in HEIs and UILs, all the three stages of technological change—invention, innovation, diffusion—get affected. Thus, building on the discussion of the three frameworks discussed in the last chapter—NIS, triple helix and UILs—policy instruments relating to education and S&T directly affect academia and research institutions.

At a broader level, as an important part of the NIS, HEIs can potentially provide a common and possibly neutral platform for discussion about the broader goals of innovation policy and a forum where there can be relatively open interaction between industry and government. More specifically, educational institutions can contribute to the innovation process through a variety of linkages (Basant and Chandra 2007a; Basant and Mukhopadhyay 2010):[*]

i. *Supply of skilled persons for innovation-related activities.* As the higher education sector evolves, it acquires the capacity to handle more complex technologies, and its influence extends to more industries. HEIs supply skills to create and apply knowledge. Given our broad definition of innovation, this 'labour market link' is not only restricted to developing new technologies but also to adapting and adopting innovations in the 3 Ps developed elsewhere.

ii. *Creation of Knowledge* through their own research or through industry-institution R&D projects. Research can also be financed by the government. The knowledge developed by the university can be licensed to external entities or this scientific and technical knowledge can spill over through publications, etc., and enhance the efficiency of R&D undertaken by other private or public organizations.

iii. *Creation of New Enterprises* whereby technologies developed in the institutions by either students or faculty are commercialized through new enterprises. Often this is done through facilities like science parks, innovation

---

[*] Also see Perkmann et al. (2013) for a comprehensive review of literature on university-industry linkages. Mowery and Sampat (2005) provide a detailed account of the role of universities in the national innovation system.

centres, incubators and accelerators that are created around research-based academic institutions to facilitate this process. Often these facilities are supported by the government.

iv. *Supply Services that contribute to innovation.* These include services like testing, training, certification, prototype development, etc.

The focus of the UIL literature has been on the creation of knowledge, its commercialization and the creation of enterprises. However, while the former types of links are possibly more important in countries where growth is driven by pushing the technological frontier, the labour market links are likely to be more significant in developing countries, where the growth of innovation activities would depend more on how quickly existing technologies can be exploited. In many industries this ability may well be constrained by the supply of highly skilled labour, which will determine the cost and the pace of expansion.

The Bayh–Dole Act in the US is one of the most researched policies in this genre. This act allowed US universities to own the intellectual property generated through research funded by the state. There is, however, a lot of scepticism about the efficacy of such an act in the context of developing countries which are very different from the US in terms of the institutional context of HEIs. Besides, the evidence on the impact of the act in the US is also mixed (So et al. 2008). It has also been argued that the impact of any act of this kind or of changes in state funding for R&D would depend on the autonomy of the HEI, its governance and the competition it faces for research funding. Evidently, if state universities receive a positive funding shock, they are likely to produce more research output (patents) if they are more autonomous and face more competition from private research universities (Aghion et al. 2009). The importance of autonomy is not only restricted to the efficacy of state research funding and policies like the Bayh–Dole Act, but it is also quite critical for their ability to determine the kind of training and teaching that they will undertake. Since labour market linkages are the most dominant UILs, a failure of the universities to develop the right skills and capabilities through appropriate courses would result in a significant capability and knowledge gap in the economy.

3.4.5. Incubation of Innovation-Driven Start-ups in HEIs

Building on the idea of NSI and triple helix, researchers and policymakers have often seen HEIs as anchors for regional or cluster development wherein the activities undertaken by the university contribute to the local capacity-building (Youtie and Shapiro 2008). This is done through all four linkages specified above. Incubators, science parks and innovation centres located in and around HEIs are considered to be promising policy tools that support innovation-driven entrepreneurship. Among these mechanisms, Technology Business Incubators (TBIs), which typically also organize accelerator programmes, are probably the most common mechanisms of technology commercialization across nations.* Different types of support to TBIs are used as policy tools with the presumption that such

---

* TBIs take different forms—technology/business incubators, innovation/technology centres, science/technology/research parks, and business/seed accelerators. The terminology typically reflects the scope of activities undertaken by the TBIs (Mian, Lamine and Fayolle 2016).

organizations provide an institutional structure to take care of extant market failures and provide critical inputs for the formation of innovation-driven firms.* These inputs to start-ups take the form of various linkages providing them access to business and professional services, exchange of knowledge between stakeholders, networking opportunities, resources of HEIs, mentoring and capital. Effectively, therefore, incubation and acceleration is 'an interactive process often involving inter-organizational collaboration between government, universities, industry and end user stakeholders' (McAdam, Miller and McAdam 2016: 69).

Invoking the NIS and triple-helix frameworks, university-based TBIs can be viewed as public-private or public-private-academia partnerships wherein different arms of the government, universities and industry collaborate to facilitate innovation-driven entrepreneurial growth. Through such partnerships, incubators based in HEIs 'connect science, technology, education, knowledge, entrepreneurial talent and capital' (McAdam, Miller and McAdam 2016: 70). Recent reviews of research have shown that the world over, the models of TBIs have evolved and have varied a great deal across nations as these partnerships have taken multiple forms. The efficacy of these models depends on a large number of factors.† But the role of public policy in TBIs is

---

* Not surprisingly, two of the many theoretical lenses used by researchers to analyse TBIs are *market failure* and *institutional theory*. According to these frameworks, incubators/accelerators address perceived market imperfections arising out of inefficient allocation of resources; the support mechanisms provide a more structured approach to reduce uncertainty and risk (Mian, Lamine and Fayolle 2016).

† See, for some interesting insights, Lockett et al., (2005) and the special issue of *Research Policy* (34, 2005) on spin-offs from

still an under-researched area (Mian, Lamine and Fayolle 2016). Among others, three interrelated features of TBIs have significant relevance for policy, especially in the context of developing nations:

i.   It is being increasingly recognized that incubation practices and models across TBIs located in HEIs should not be driven any more by the 'global best practices' but should explicitly recognize the unique features of the region of their location and characteristics of the host HEI (Basant and Cooper 2016; McAdam, Miller and McAdam 2016);

ii. Policy support for TBIs does not pay much attention to the fact that incubators' ability to provide specialized services to start-ups is probably more critical for their success than physical infrastructure. Consequently, resources for developing or acquiring skills (especially soft ones) are typically not available to TBIs, especially those located in developing economies (Sharma and Vohra 2020); and

iii. Government support to TBIs in HEIs also does not have a clear understanding of how the TBI would achieve financial viability in different regional and institutional contexts. The support usually takes the form of seed support for setting up TBIs but its financial and organizational sustainability in the future is not explicitly considered (Basant and Cooper 2016).

Due to these and other reasons, TBIs across the world are trying out incubation and/or acceleration models that make sense in their regional and institutional contexts (Mian, Lamine and Fayolle 2016). Besides, policymakers are also experimenting with various policy options. These take the

---

public research institutions.

form of linking funding to certain outcomes, for example, the ability of the HEI to actively engage with industry and end users as a part of the incubation process (McAdam, Miller and McAdam 2016). At times, the government funding for innovation and incubation is linked to a number of HEIs collaborating in the endeavour and complementing each other's efforts and capabilities. This 'forced collaboration' model is expected to create an innovation ecosystem wherein a cluster of HEIs complement each other to fill in gaps in capabilities required to innovate specific technology domains. SETsquared partnership is one such funding initiative in the UK that endeavours to bring together five universities in a region to collaborate on innovation and entrepreneurship.*

In recent years, the governments in most developing nations, including India, are supporting the creation of incubators in HEIs to facilitate commercialization of university and external inventions through new enterprise creation. The next chapter will discuss some of these issues in the context of India.

3.4.6. *Intellectual Property Rights (IPR) Policy*

As the discussion earlier in the chapter would suggest, the links between the IP regime and innovation activity are quite complex. IPRs affect competition and contagion conditions in several ways. More stringent IPR policies provide incentives to innovate due to increase in appropriability (monopoly rights). This in turn might have a positive impact on firms' R&D expenditures and patenting activity—innovation input

---

* For details, see https://www.setsquared.co.uk/

as well as output.* Clearly defined IPRs facilitate innovation-based entry by smaller and new firms. Moreover, well-defined rights on intellectual property also create a market for technology as they reduce transaction costs and thereby create a potential for increase in competition through purchased innovation-based entry. With stringent IPRs, the price of technology may be high due to monopoly over the technology but it is possible that in a competitive market it may not be exorbitant. Moreover, the owners of technology may now charge a lower 'risk premium' as the imitation potential is lower with stringent IPRs. If incentives associated with stringent IP policies result in more knowledge getting *generated*, *introduced in the market* and *transferred through the market* due to the development of technology markets, economic entities would have a larger knowledge pool to learn from. Consequently, *aggregate* spillover potential might increase. With a better IPR regime, more R&D and patenting by domestic firms would add to the 'spillover stock (or pool)'. MNCs may also do more R&D and manufacturing in high-tech sectors, especially in those where imitation potential is high (e.g., chemicals, pharmaceuticals) with more stringent IPRs. Besides, typically more 'masking' of knowledge is done in a weak IPR regime in order to reduce imitation which reduces contagion potential; in a more stringent regime such masking may get reduced and more tacit knowledge may be shared by the IP owner.

While it is not possible to attribute all types of innovations to changes in the IP regime, there is evidence of a variety of innovations in the health sector in India after the introduction

---

* In a cross-national empirical study of OECD nations, Wu, Popp and Bretschneider (2007) have found that patent protection stimulates business R&D.

of a more stringent TRIPS-compatible regime (Basant and Srinivasan 2016). As expected, another study on India suggests that the impact of such reforms is more in 'IP-sensitive industries' like non-electrical machinery and drugs and pharmaceuticals than in other industries (Kanwar 2013). In line with the insights of Cohen and Levinthal (1989), a firm-level study shows that while the use of IPRs (patenting intensity) reduces competition in the market, it also increases innovative activities (R&D expenditures, product innovations) (Beneito et al. 2014). In fact, some studies have suggested that an *intermediate* position, in terms of the tightness of the appropriability regime, may be most effective for a firm as well, as it may provide more control and alternatives to react to emerging opportunities (Hurmelinna, Kyläheiko and Jauhianinen 2007). It is not clear if stringency of IP regimes and innovative activity have a linear relationship; many argue that the most stringent IP protection may not be optimal for the economy, given the regime's simultaneous and differential impact on invention, innovation and diffusion (Ordover 1991).* As discussed, very stringent IP regimes may encourage inventive and innovative activity but diffusion may be adversely affected. The impact of IP regimes on innovation is also mediated by levels of development (Falvey and Foster 2006). Moreover, very little is known about the implications of varying degrees of IP protection on the make-buy-copy combinations.

---

* Reitzig (2004) strongly highlights the anti-competitive effects that arise out of a patenting system wherein multiple patents are granted per innovation in complex and discrete technologies. Given degrees of technological complementarities, patentees could eliminate competition in the form of substitute technologies through such fencing.

It has been argued, therefore, that policymakers should analyse general appropriability conditions to understand why the ability to use IPRs and their efficacy varies for different types of firms instead of assuming that a stringent IP regime is the best IP solution.

3.4.7. Policies for Technology Licensing

Technology licensing policy takes the form of government intervention in markets where private enterprises acquire technology. While it is not very common any more, earlier several governments used to regulate the licensing of foreign technology (Steinmueller 2010). Such regulation affects the extent, price and vintage of technology that gets licensed. In India, for example, during the pre-1991 phase, it was extremely difficult to license foreign technology. Such a request not only had to satisfy the indigenous non-availability requirement but also had to abide by very stringent restrictions on the royalty that could be paid and the conditions of technology transfer. Even in the absence of such restrictions, this policy has implications for the extent and nature of knowledge transfer through the licensing process. This is especially so if the policy is looked at in conjunction with the IPR policy. A liberal licensing policy combined with a stringent IP policy may facilitate transfer of knowledge, especially tacit, through licensing-linked training. This in turn has a positive impact on the contagion potential. Overall, this policy affects potential entry (through licensing) based competition as well as competition among incumbents as they can also license technology and compete. The pool of knowledge that is transferred generates potential for contagion or knowledge spillovers. Since technology licensing policy directly affects technology purchase options (especially from foreign firms), this policy would influence knowledge-sourcing strategies of firms (Basant and Fikkert 1996). But hardly anything is known about how these policies change firm preferences with respect to make-buy-copy combinations to acquire knowledge (Basant 1993).

3.4.8. Policies for Standard Creation

Standards affect demand and supply side network economies in the relevant market. Policy can allow standards that are based on market competition where multiple standards coexist, it can create standards and make market entities follow it or it can be neutral vis-à-vis standards but ensure inter-connectivity of standards. There is innovation to meet standards and, as creation of standards typically increases the size of the market especially through network economies, there is a higher supply of innovation to satisfy the demand of the growing market. With firming up of standards, there can also be a focus on innovations in standards compatible complementary technologies that can help enterprises enter new markets (Jacobides et al. 2006; Shin et al. 2015; Blind 2013; Allen and Sriram 2000). In the context of India, a recent and probably one of the most critical policies related to telecom standards. By being standards neutral and ensuring inter-connectivity, the Government of India not only reduced regulatory uncertainty for telecom players and enlarged the market, it also did not take a bet on a specific technology standard. This way they managed the uncertainty about which standard would become dominant and would be more efficient.

3.4.9. Public Procurement, R&D Funding and Tax Credits

Policies affecting *procurement, funding* and *cost of capital* also have an impact on innovation activities. The state can procure 'innovative' products on a preferential basis from specific types of firms like SMEs or start-ups. Such support can also take the form of adoption subsidies for using new technology-based products or services in order to facilitate diffusion of new technologies (Steinmueller 2010). The state can directly finance an innovation activity as well, which can take the form of 'thematic funding' to target and take advantage of opportunities that are specific to a technology, sector and even a region. Such funding is more common for pre-competitive research.* These policies can support innovation but can also substitute or complement private spending on R&D by enterprises. While venture capital may help innovation-based entry by start-ups, availability of debt funding and its cost can influence purchase of embodied innovation (e.g., new machines which can also act as collateral). Consequently, policies that directly or indirectly influence procurement of innovative products, funding of innovative start-ups and cost and availability of debt capital for the acquisition of embodied knowledge have implications for make-buy-copy choices of economic agents.

Public R&D and provision of subsidy or tax credits for R&D enhances R&D activity† which in turn increases the

---

* See Steinmueller, 2010 for a discussion on such support.

† For some evidence on the positive effect on business R&D of R&D performed by the government, R&D tax incentives and government funding of private R&D in OECD nations, see Wu, Popp and Bretschneider (2007).

knowledge pool (contagion potential) as well as possibility of innovation. At the margin, it will also impact the composition of make-buy-imitate choices among the private entities, with the 'make' option becoming more preferred, *ceteris paribus*. Available literature suggests that the type of R&D support may have different implications for innovation-related activities. In a developing country, public research is more likely to focus on basic research and early-stage technology development while private research (with or without tax subsidy) would typically give priority to late-stage technology development. Private sector research may also focus relatively more on learning from others, benefiting from global commons, building absorptive capacity and undertaking context-specific adaptive innovation. The empirical evidence broadly suggests that tax incentives positively influence R&D spending of firms which are *already* engaged in such activity but an unequivocal assessment in this regard is difficult (Guceri 2016).

Whether R&D subsidy substitutes or complements firms' own efforts is an under-researched area. A recent paper on the effect of R&D subsidies on Chinese medium and large firms shows that up to a threshold such subsidies enhance firms' innovativeness; beyond this threshold the subsidies seem to substitute firms' own efforts. Of course, as one would expect, a firm's own R&D capacity, its size and industry technology levels affect the efficiency of R&D subsidy (Zhang and Wu 2014). Broadly, while a few recent studies suggest that R&D subsidy complements firm-level R&D, some crowding-out cannot be ruled out (Busom 2000; Zhang and Wu 2014). Given this, the policy choice between R&D subsidies (in the form of direct research grants) and R&D tax incentives is difficult. While the latter is often preferred for interfering less with the market mechanisms as firms decide the focus of

R&D efforts, the former may be required to focus on research in areas where market failures are high.

While the relative efficacy of different mechanisms for public support for R&D (e.g., grants vs tax incentives) is still somewhat unknown, the fact that well-directed public R&D support is useful is reasonably well established. As mentioned, research grants are seen by some as 'interfering' with market mechanisms and therefore tax incentives are less distorting. Besides, in choosing areas/projects to be supported through research grants, governments may also fail. But the fact remains that left to themselves, private entities would choose research projects that may provide quick returns and keep away from areas with positive externalities and from projects that might only provide long-term returns. Therefore, the role of government in supporting innovation in such areas cannot be denied. In fact, there is ample evidence to show that several programmes providing public funding for R&D have been quite successful. In fact, public funding has been useful in the entire innovation supply chain. For example, in the US, the National Science Foundation (NSF) has been critical for basic research while institutions like Defence Advanced Research Projects Agency (DARPA) have supported transnational institutions in a big way. Moreover, Small Business Innovation Research (SBIR) has successfully provided long finance for research in private enterprises. Similarly, sustained support by National Institutes of Health (NIH) for health-related research has led to significant advances in biotech and development of several innovative drugs. Provision of patient capital for research is not only available in the US. Most other countries provide such support, although the design of support varies. Many countries, for example, work through state investment banks (SIBs).* The key issue in all these interventions is to ensure that state support takes care of market failures but uses processes of selection, etc., which minimize the chances of government failures. PPPs can be useful in this regard and if the state is able to create a vibrant coalition of different stakeholders in the NIS—government, business enterprises, academia and others—such partnerships co-shape markets rather than only fixing market failures; and such coalitions can be very useful in situations wherein the state is seeking innovative solutions to wicked societal problems through a mission mode (Mazzucato 2018).

3.4.10. More on Financing Innovation and Cost of Capital

Innovation, especially the new-to-the-market variety, is expensive, uncertain and takes time, requiring patient capital. A well-functioning financial market can spur technological innovation by allocating resources to enterprises with a higher chance of introducing innovations or by financing certain innovation-related activities directly. The latter can be done by supporting R&D and facilitating commercialization of new technologies. A few characteristics of innovative efforts (like R&D) and the innovation process contribute to imperfections in the market for financing such activities†:

i. Innovation activities are inherently uncertain with unknown potential outcomes. This 'fundamental

---

* For basic information on some of these programmes, see Mazzucato (2018).
† For a detailed discussion, see Kerr and Nanda (2015) and O'Sullivan (2005).

uncertainty' is critical for financing decisions that are close to the technology frontier. Other adaptive innovation efforts can also be quite risky.

ii. Returns from such innovation processes are very skewed and when combined with high uncertainty, make the standard ways of evaluating projects inadequate. Specialized entities with their own heuristics are often required to judge the value of such projects.

iii. While neither the innovator nor the financier may be fully aware of the potential of an innovation project, the innovator may still know more, resulting in agency costs associated with information asymmetry. Drafting near complete contracts in such situations with appropriate contingency clauses becomes almost impossible as outputs at times are unknown.

iv. Firms actively engaged in innovation activities have a significant share of their knowledge base that is tacit and this is primarily embedded in the human capital of the firms' knowledge workers. It is critical for the firm to retain such employees and therefore, they have to maintain a threshold (usually high) level of R&D spending at all times irrespective of the intensity of innovation activity. This requires an assured and constant source of funding. Additional expenditures may be episodic depending on the needs of the research portfolio.

v. External financing of innovation, especially in the early stages when IP protection is inadequate, involves sharing of critical information by the inventor with the financier. This might compromise or reduce appropriability for the innovator and s/he may prefer internal funding.

Given all these issues, firms may have to rely on internal funds for much of their innovation-related investments.

The R&D subsidies and tax credits discussed above can facilitate internal financing as these policy instruments affect cash flow of firms. There is evidence to show that relatively young, high-tech and publicly listed firms in some developed economies typically finance their R&D investments from internal resources and external equity markets (Kerr and Nanda 2015). However, there is recent evidence to suggest that the role of debt capital in financing innovation is on the rise, at least in developed economies. Typically, patents are used as collateral for raising debt capital including in new venture situations. While bank finance is important for innovation, it is likely to be more prevalent for larger firms, where loans are less risky for banks, given collateral options of fixed assets apart from intellectual property.*

Financial constraints, however, may not be of the same severity for all kinds of innovation activities as the degrees of uncertainty may vary across projects. Adaptation and adoption projects are less uncertain as technological uncertainties are much lower and the financier needs to worry essentially about the market risk. As mentioned, availability of debt capital can be quite critical for adoption of embodied technologies. Therefore, innovation that is new to the firm but not new to the market requires financial resources and typically this will take the form of internal capital or external debt capital. In cases where a firm develops and commercializes an innovation, resources need to be invested in the learning process as well as commercialization that would require getting access to complementary assets needed for taking the innovation to the market. Insofar as the learning process requires specialized skills and capabilities, public financing of

---

* Kerr and Nanda (2015) provide a brief review of these studies.

education (like public sector R&D mentioned above) can also be viewed as financing innovation.

In general, raising money for R&D projects from the financial system is quite difficult and at times gets raised through grants, both from public and private sources like foundations. Apart from the nature of innovation (degrees of uncertainty and risk), the scale and scope of the firm may affect availability of finance for innovation. The cost of raising capital is important both for developing a technology as well as commercializing it, given the need for the access to complementary assets. Consequently, high cost of capital in general can dampen innovative activity. As mentioned, larger, publicly listed firms may be able to raise debt and equity capital more easily than smaller firms and start-ups. Valuation of patents in developing economies as collateral for providing debt capital for small firms is likely to be difficult given the absence of specialized skills. Besides, very few start-ups and small firms would have patents with commercial potential to leverage such opportunities. Therefore, regulations and policies that affect availability of capital for smaller firms through listing requirements in stock exchanges and availability of debt and equity capital through other sources are critical for innovation in such firms.

In situations of the kind described here, the role of VC is quite well known in most developed nations (Lerner and Tag 2013). VCs are expected to address the capital market imperfections enumerated above and use their own heuristics to evaluate innovation-driven new enterprises and provide early-stage funding. An entrepreneurial ecosystem that includes robust angel investment opportunities helps the availability of finance even further as VCs do not need to invest at a very early stage. In the same vein, availability of seed capital that many incubators and accelerators provide, often with the support of the government agencies, reduces the cost of experimentation for the VCs, as the innovation-driven start-ups have already come through early, riskier stages. In that sense, creation of a vibrant innovation and entrepreneurial ecosystem through various policy instruments can also reduce the constraints on financing innovation activities as well as innovation-driven ventures, both new and old.

3.4.11. Industrial Clusters Policy

In many parts of the world, governments have used a variety of policy instruments to either *create* industrial clusters or *support* market-induced clusters.* These instruments take the form of investments in infrastructure, free-trade zones, SEZs, setting up of cluster-specific common facilities for testing, etc., and even establishing HEIs in the cluster. Physical agglomeration of diverse firms and institutions along with coexistence of competition and collaboration in these clusters creates both competition and contagion effects. Flows of knowledge take place in these locations through various mechanisms including participation in intra-cluster or global

---

* At times, clusters emerge partly in response to policy initiatives and become very efficient with the help of significant inter-firm linkages and division of labour. But they face problems when the policy is withdrawn or the environment changes as they get locked into an obsolete technology supported by the policy. See Basant (2002) for discussion of the diesel engine cluster in Rajkot, India which flourished during policy regimes that subsidized low-speed, low-HP diesel engines. Changes in these policies and the emergence of high-speed diesel engines of Chinese origin posed significant problems for Rajkot firms, specializing in different components and sub-assemblies, as they did not have the capability to easily move into the new technologies.

production or innovation networks (Basant 2002; Basant and Chandra 2007; Yusuf, Nabeshima, Yamashita 2008). While firms in industrial clusters are found to be more technologically active and efficient as compared to non-cluster firms, there is also a possibility of firms in clusters getting locked into specific products and technologies. It has been argued that clusters where firms face competition (existing as well as potential) and are part of 'internal' and 'external' networks are more sustainable and technologically active (Humphrey and Schmitz 2002; Iammarino and McCann 2006).

Industrial cluster-related policies often result in the creation of innovative regions like Silicon Valley. Such clusters can be seen as mini innovation systems wherein different actors identified as parts of the NIS are very active locally. Since knowledge is often sticky and difficult to transfer due to its complexity and tacitness, co-location of different actors facilitates knowledge flows. Vibrant cities often become the loci for the emergence for such innovative clusters. Insofar as policies can facilitate the growth of such clusters through investments in activities that can create opportunities for multiple actors to interface and provide core innovation-related infrastructure, such clusters can be an important component of the innovation policy.

3.4.12. Policies for Information and Communications Technology (ICT)

The processes relating to innovation are typically iterative. Designs, prototypes and ideas are often shared with people with different skills and perspectives. The emergence of ICTs has created new opportunities for building such connections across boundaries into a digital world and undertaking iterative interactions critical for innovation outcomes and their diffusion.

Digital technologies, in fact, have become the core of innovation-related technology infrastructure. Such technologies combine design and manufacture through computer-aided design and manufacturing. Digital systems can help develop designs that are more manufacturable, making innovation activity more efficient. Organizations, in fact, have experienced significant changes in the way they function and undertake innovation-related activities. Consequently, digital technologies that ride on the ICT infrastructure have implications for all types of innovation—product, process and practice.

Developments in computing power, software and new technologies like AI, ML and others relating to data analytics have given rise to a new kind of technology supporting innovation which Dodgson and Gann (2010) refer to as 'innovation technology'. Pervasive use of such technologies has affected innovation processes in significant ways. It has changed prototyping methods as well as mechanisms to get user inputs. These technologies permit organizations to experiment cheaply and fail early.*

Overall, ICT adoption is an innovation in itself for a firm as it is the use of new technology at the firm level. At the same time, intensive and diverse use of digital technologies that ride the ICT infrastructure can change innovation processes in myriad ways. In fact, these technologies can potentially aid integration of the 3 Ps. Given this, policies that affect the availability of ICT infrastructure and access to digital technologies can be very important for innovation activity and outcomes.

---

* See Dodgson and Gann (2010: pp. 113–14), for a succinct summary of ways in which innovation activity is changing due to the emergence of ICT and digital technologies.

Many of the policies discussed above can also be seen as initiatives that make institutional changes to facilitate innovation. These include policies relating to UILs, education, ICT infrastructure, industrial clusters, thematic funding for research in the public and private sectors, support for technology cooperation, regulations for financial sector markets, standards and so on. Such institutional changes help build national and regional systems of innovation—a critical set of informal and formal institutions which help build innovation capability. As discussed in the last chapter, the literature using the systems of innovation approach also emphasizes iterative learning (both learning by doing and learning by using) and non-linear relationships between processes of invention, innovation and diffusion (Fagerberg and Sapprasert 2011; Steinmueller 2010).

3.4.13. Interaction between Policies

The brief discussion on the role of various policies suggests that conceptualization of *direct* policy effect is quite complex. If one introduces interactions between policies (some of which have been mentioned above), the analysis becomes even more challenging. Few studies have explored these interactions and it would be instructive to mention some of them to bring out the complexity.

There is some evidence to suggest that liberal trade policy results in more flows of embodied knowledge through high technology imports when IP regime is stringent (Briggs 2013). It has also been suggested that transfer of embodied knowledge through trade as a result of changes in IP policies is mediated by education levels of the workforce, absorptive capacity, public funding of R&D, quality of infrastructure, market structure and other government regulations (Akkoyunlu 2013).

This obviously makes the policy interactions really complex. In the same vein, and as partly alluded to above, the nature of FDI and the associated competition and contagion effects are conditioned by IP regimes, absorptive capacity of firms (which is dependent on firms' own technological activity), licensing policy and trade regimes. There is also some evidence to show that stronger IPRs: (a) increase technology licensing, especially in countries with higher imitative/innovative capacity; (b) result in technology-intensive trade and associated spillovers in countries with larger markets and higher imitative capacity; and (c) lead to higher foreign patenting (and spillovers) in more open economies and in those with higher innovative capacity[*] (Falvey and Foster 2006; Akkoyunlu 2013).

Moreover, FDI-related spillovers are also affected by the institutional environment—corruption, red tape, level of development—in the host country (Gorodnichenko, Svejnar and Terrell 2014). In line with the arguments above, a detailed firm-level analysis of US MNCs by Branstetter, Fisman and Foley (2006) shows that stronger IP regimes increase the technology transferred to MNC affiliates, and R&D expenditures of affiliates as well as the level of foreign patent applications. This implies that a liberal FDI policy combined with a stringent IP regime can result in better technology transfer and also create a larger potential of knowledge spillovers as MNCs bring in more technology to the host economies. Similarly, R&D subsidy and support is typically found to be more efficacious in open economies (Falvey and Foster 2006), suggesting R&D and trade policy

---

[*] Interestingly, there is some evidence to show that better appropriability through effective patent strength and secrecy facilitates technology licensing process from HEIs (Dechenaux et al. 2008).

interface and such subsidization works better when firms face import competition.

In the context of technology licensing, Arora (1991) has argued that *codified* elements of the technology are more valuable when used in conjunction with the *tacit* know-how. This is so because both are strongly influenced by the *specific conditions* under which the licensor operates the technology and learns about it. Also, the stronger the complementarity between the codified and the tacit components of the technology (i.e., higher the degree of tacitness of the technology), the broader the scope of protection afforded to the codified elements of the technology. Therefore, under a more stringent IPR regime, the cost incurred by the licensee to 'invent around' the licensor's technology is likely to be more. Consequently, *ceteris paribus*, the amount of tacit knowledge provided in a licensing arrangement may decline with the weakening of IPRs. Thus, if a country adopts a weak patent regime, the possibilities are that the licensors will supply lower levels of technological know-how by restricting the flow of tacit knowledge.

Competition policy is evidently critical for countering the ill effects of IP-based monopoly—both in terms of contagion and competition effects (Globerman 2012; Dumont and Holmes 2002). It is also argued that the cost of capital disadvantages may result in domestic firms not being able to meet the innovation-based competition from imports and FDI, which reflects the interaction between the financial sector, trade and FDI policies.

In a theoretical paper, Mohnen and Roller (2005) add an interesting insight into the issue of complementarity between policies to support innovation. They argue that there is a need to adopt a *package* of policies to enhance the *propensity* of firms to innovate, but for improving the *intensity* of innovation, a more *targeted* choice of policies is required.

In other words, the complementarity of policies varies for different phases that are being targeted. In an empirical exploration of this model, Strube and Resende (2009) find some supporting evidence in the context of Brazilian firms. In the same vein, Westmore (2013) also argues that while R&D tax incentives, patent rights and direct government support encourage innovative activities, policies of product market regulation, openness to trade, debtor protection and bankruptcy provision are important for innovation and diffusion of technology.

One can multiply such examples to show that analysis of policy-innovation linkages becomes quite complex when we consider a plethora of policies that can influence innovation and simultaneously explore the effects of the interactions among various policies. The challenge for the policymakers in the globalizing world of open and distributed knowledge networks is the need to identify a policy package that can simultaneously facilitate international linkages for accessing knowledge, incentivize domestic intra-mural R&D to build absorptive as well as inventive capacity and help create domestic networks for knowledge accumulation and diffusion (Herstad et al. 2010). It is well recognized that the most appropriate policy package is one that encourages innovation as well as knowledge spillovers to become pervasive. The elements of this 'package' are, however, not easy to unravel (Westmore 2013). Unfortunately, there are hardly any studies which rigorously analyse the impact of various policies on innovation in India, leave aside those which explore the interactions between policies.

3.4.14. The Measurement Challenge

It has been argued that the relationship between innovation measurement and innovation policy needs to be appreciated;

such an appreciation helps devise more efficacious policy instruments that recognize various dimensions of innovation and the underlying process.[*]

Before concluding the chapter, it is important to recognize the measurement-related challenges one faces in analysing linkages between innovation and public policy. The empirical studies summarized in this chapter also face them and the recognition of such challenges adds significantly to the complexity of analysing these linkages. Even when one is able to conceptually specify appropriate relationships (which is quite complex as we have seen in this chapter), a big challenge is to measure various effects. Any empirical exploration of innovation-policy linkages would require measures not only for innovation outcomes and innovation-related activities but also for various policies. This task is far from trivial.

The last chapter identified a variety of measures for innovation inputs and outcomes (R&D and technology purchase expenditures, patents, different types of innovation) and also discussed some estimates from innovation surveys. The surveys provide information on sources of ideas and information that were critical for innovation as well as data on various linkages that facilitate knowledge flows and innovation. All these, as mentioned, can be seen as measures of contagion or imitation potential. Contagion or spillovers potential is also measured by accumulated *stocks* of R&D expenditures, patents and FDI at the *industry* level, assuming that such stocks capture knowledge pools made available by innovation-related activities of the

---

[*] Gault (2016) and Sloan (2006) provide an excellent discussion on how important measurement of innovation, innovation-related activities and the process of innovation are to devise appropriate innovation policies.

economic agents and also through MNC engagement with the economy. The higher the intensity of R&D, MNC activity and patenting in a particular industry, the higher the expected imitation or learning potential. Often such stock measures distinguish between domestic and foreign R&D and patents to develop spillover stocks emanating out of foreign and domestic knowledge generation. The role of stocks of purchased technology (innovation), and even training, has been typically ignored in such estimates but can be quite important in providing avenues for reverse engineering and in absorbing technology. Most studies use aggregate FDI stocks in different industry groups to capture knowledge spillover potential of foreign investment. At times these stocks are weighted by R&D, technology licensing or patenting done by the MNCs to make them more innovation-focused.

Innovation/imitation/absorptive capacity is also usually measured through patenting and R&D expenditures. This is typically done at the firm level but is measured at the industry, region or country level as well, depending on the analytical needs. At the regional or national level, and at times at the firm level, standardized number of skilled workers (e.g., scientists, engineers) is also used as a measure of such capacity. As mentioned above, innovation surveys are used to capture innovations in the 3 Ps (products, processes and practices), purchase (adoption) of innovations and to some extent, contagion effects and innovation capacity. Large surveys are very few and are difficult to organize. As seen in the last chapter, all these measures are quite incomplete in many ways and the conclusions based on them need to be used with caution.

Measuring policies and changes in them is equally complex. For various policies discussed in the chapter, several measures have been used empirically to capture *differences* in policies *across regions* or *change* in them *over time*. For example, IP policy indices that capture stringency of IP regimes across nations are very commonly used. Very few of these, however, capture enforcement or implementation issues. Within country analyses, *before* and *after* exercises are typically undertaken to capture policy impact. For example, many studies on the role of IP regimes in India focus on situations before and after 2005 when the new regime was put in place. Similarly, studies on the impact of R&D subsidy/tax policies also usually analyse *before* and *after* situations. There are very few cross-country studies as they require comparable indices (Gault 2016). Trade policy is usually captured through openness indices and various other measures to capture tariff and non-tariff barriers at the level of industry groups. These measures are too aggregative at times to be meaningful at the firm level. FDI policy impact is generally not captured in innovation studies per se but as mentioned above, 'contagion potential' is measured with aggregation of *FDI flows* over a specified period or *FDI stock* in an industry group at a point in time. Typically, periods of FDI liberalization are identified to measure policy impact but sectoral and sub-sectoral differences are often difficult to capture as detailed information is not available.

Studies have also not been able to measure industry policy liberalization adequately to capture entry and exit conditions and degrees of contestability. Usually, liberalization periods are identified for specific country studies and these are combined with measures of competition like concentration and price-cost margins. However, these measures not only capture the impact of policy but also that of industry characteristics and strategies of firms in those sectors, and therefore cannot be seen as adequate in all respects. Cross-country indices are only developed for overall liberalization and capture all policies; they are not able to capture regional or industry-specific variations. Some studies use self-reported measures of the competition that firms face in a market.*

Costs of capital measures are available but they are typically not linked to innovation. Here again, these are only partly determined by policy. However, a few recent studies explore the impact of VC funding for innovation-driven start-ups. Interestingly, here again, the ability of venture capital to be efficacious is found to be in situations where several legal and other institutions are in place along with appropriate labour market and tax regulations.†

Similarly, several studies (mainly theoretical) have analysed 'optimal patent policy design' but empirical work has not provided consistent results. Measurement issues may have resulted in such variations in results. Measuring patent strength, for example, is quite a non-trivial and difficult task. It has been argued, therefore, that any exploration of the relationship between patent strength and research investments needs to address the challenge of measuring specific research investments and variations in degrees of patent protection (Williams 2016).

It needs to be emphasized that in the absence of appropriate measures, it is difficult to undertake cross-country analyses of policy-innovation linkages and explore the impact of interactions between policies within countries as well.

---

* As mentioned in the introduction, measures of global innovation index also attempt to measure cross-national differences in policies.
† For some useful insights on this, see Lerner and Tag (2013).

3.5. Concluding Observations

The chapter highlights the conceptual and empirical complexities that a policymaker faces while analysing the linkages between public policy and innovation. The theoretical and empirical studies on the impact of different policies on innovation are mixed and somewhat difficult to interpret, given these conceptual and measurement issues. So, what can one take away from the available literature and data? Can one identify salient areas of intervention for innovation-driven long-term productivity growth?

Despite all the analytical complexities, one of the most consistent results in the available literature is that *all* policies that can potentially affect innovation-related outcomes—R&D, technology licensing, IPRs, industry, trade, FDI—tend to have a positive impact in situations where innovation capability and absorptive capacity is reasonably high. Also, innovation policy instruments are more efficacious if levels of competition and openness are high. These results along with the earlier ones strongly suggest that the success of the liberalization process will crucially depend on the quantum and nature of investments Indian firms make in building their innovation or technological capabilities. The major policy challenge, therefore, is to identify instruments which induce such investments without curtailing the flow of technologies from abroad. It is not clear from available studies if liberalization since 1991, FDI inflows, spurt in foreign collaborations, joint ventures and marketing tie-ups are combined with serious technology-related investments by Indian firms to attain technological dynamism and competitiveness in the long run. We analyse some of these trends in the next chapter and also discuss some new innovation policy instruments that have been used in India in recent times.

4

Innovation and Innovation Policy in India

A Focus on New Initiatives

4.1. Introduction

In the last chapter, we discussed the broad conceptualization of innovation policy, defining it as a set of policy instruments that affect the choices of economic agents with respect to innovation efforts and outcomes. The complexity of analysing the linkages between public policy and innovation was also discussed in some detail. It was highlighted that there exist very few studies on India that analyse a variety of relationships explored in other countries, presumably because of the non-availability of appropriate data. Of course, some studies on India were referred to in the chapter and some data was shared. This chapter focuses more sharply on India. Since there are few studies on India and the available data leaves much to be desired, the chapter provides insights from available studies, attempts a preliminary empirical exercise to explore the aggregate trends in innovation in India during the post-liberalization period and reviews some new innovation policy initiatives.

Several studies have analysed the policy changes that have been introduced in India since 1991 and their impact on the economy. Without getting into the details of the policy changes, it can be said that, barring a handful of exceptions, the movement over time has been towards liberalizing the economy on all fronts—trade, FDI, industrial policy, infrastructure, etc.—to enhance the internal as well as external competition and facilitate knowledge flows. The policies prior to initiation of economic reforms in 1991 were broadly characterized by multiple controls over private investment, both domestic and foreign, and trade, especially imports. It is well recognized that policy rigidities of the pre-reform era adversely impacted the economy and made enterprises less focused on innovation.* In contrast, the new policy regime focuses on more competition and openness to attain international competitiveness, develop indigenous capacity in technology and manufacturing, widen capital markets, encourage foreign investment and technology collaboration and curtail monopolistic tendencies.† Some other policy measures include greater emphasis on knowledge-based industry and export of services, setting up special economic zones (SEZs), enactment of the Foreign Exchange Management Act (FEMA), attracting FDI through automatic approval route of the RBI and private sector participation in infrastructure (e.g., multiple operators in telecom sector) and insurance sectors. As in the case of FDI and embodied technology in the form of physical imports,

---

* For example, rigidities in investment policy constrained firm-level choices and limited competition leading to inefficiency, whereas industrial licensing and product reservation for the small-scale sector inhibited firms from reaping economies of scale.

† Basant (2000) provides a detailed discussion on various policy measures introduced in the early 1990s and their implications for the Indian corporate sector.

disembodied technology imports were also liberalized making technology licensing from abroad easier. Restrictions on what types of technology could be licensed and royalty rates that could be paid have been eased.*

In addition, there have been many important changes in the regulatory structure. For example, India's obligation to sign the TRIPS agreement in 1994 has caused amendments to the Indian Patent Act (1970) resulting in a marked shift from the process patent regime in certain sectors towards an era of product patents, particularly for pharmaceuticals and food products. The term of patent has also been extended to twenty years. Similarly, following the wave of M&As during the first decade of economic reforms, the Competition Act, 2002 was enacted along with the establishment of the Competition Commission to regulate M&As and to curtail unfair and restrictive trade practices as well as abuse of monopoly power.†

Contestability in the markets has been enhanced through all the policies enumerated above, especially trade liberalization, abolition of industrial licensing and adopting a liberal policy vis-à-vis FDI. While trade liberalization has enhanced firms' choices with respect to foreign purchase of

---

* After the initial reforms in the 1990s, liberalization of foreign technology agreements was again undertaken in 2009 whereby limits on royalty and licence fee (lump sum payment) have been completely removed. Prior to this change, the limit on royalty payments was 8 per cent on export sales and 5 per cent on domestic sales and in the case of lump sum payment, the limit was USD 2 billion (Mani 2020).

† As compared to the MRTP Act, the Competition Act focuses more on behaviour of enterprises and not so much on market structure. For example, the new Act makes pre-notification mandatory if the threshold value of assets of merging/acquiring firms or of the respective business groups is beyond the threshold limit.

embodied technology, liberalization of technology licensing policies has increased the choice set of firms with respect to accessing disembodied technology from foreign firms. A TRIPS-compatible IP regime has enhanced appropriability of enterprise innovations that are amenable for IP protection. A relevant question is whether these changes in the IP regime coupled with enhanced contestability have facilitated Indian firms to be more innovation-oriented and enhanced their participation in the global technology and production networks. Given the criticality of knowledge flows in a network of relationships for the transition of Indian firms to become more innovative, the impact of IP policy in building such networks would also be useful to analyse.*

Apart from the changes in the conventional policy instruments mentioned above, a large variety of innovation policy-related experiments are under way in India. These experiments relate to the non-conventional domains wherein the key idea is to create a vibrant ecosystem for innovative entrepreneurial activities. Fostering of innovation-driven start-ups is being attempted through various mechanisms. For example, incubators are being set up by different government departments in educational institutions, seed funds are being provided to develop prototypes and take inventions to the market, university-industry linkages are being supported, a Bayh–Dole-type act has been drawn up so that educational institutions can own and commercialize IP generated through state-funded research projects. Opening up of the financial markets has resulted in rapid growth of the venture capital industry, making early-stage financing for new enterprises somewhat more accessible.

---

* For a preliminary analysis of the new IPR regime for the Indian pharmaceutical and biotech sectors with the focus on global networks, see Basant (2011).

This chapter focuses on how Indian enterprises have responded to the package of liberal policies implemented during the post-1991 period. Since analysing links with specific policy changes can be very complex, as seen in the last chapter, the focus is mainly on the broad trends in innovation activities and innovation outcomes in the Indian economy. This is complemented by an analysis of policies that focus on supporting innovation-driven start-ups and on strengthening a critical element of the innovation system, namely academia.

4.2. Recent Trends in Innovation and Innovation Activity in India

Studies reviewed in the last chapter suggest that a stringent IP regime, tax and other support for R&D combined with liberalization on various fronts—trade, FDI, technology licensing, industrial investment—that enhances competition and contagion potential can positively impinge on innovation outcomes if certain conditions are satisfied. Primary among these conditions is the presence of reasonably high innovation capability and absorptive capacity among the enterprises in the economy. Therefore, the success of the liberalization process and other policy instruments would crucially depend on the quantum and nature of investments Indian firms make in building their innovation capabilities. This section analyses data on innovation-related investments by Indian firms and their outcomes.

4.2.1. Conventional Measures of Innovation: Recent Trends and Patterns

As discussed in Chapter 2, R&D, technology purchase and patents are conventional input and output measures for innovation. Pooling together data from various sources, we analyse the changes in the innovation activities and outcomes *over time*.

4.2.2. *Input Measures or the Sources of Knowledge*

Based on *aggregate* statistics, national R&D expenditures went up from about INR 3974 crore in 1990–91 to about INR 1,23,848 crore in 2018–19, almost a thirty-fold increase in current prices.* But the share of R&D in GDP (R&D intensity) increased only marginally from 0.6 per cent in 1995–96 to 0.8 per cent in 2011–12 and declined again to 0.7 per cent in the subsequent period (Figure 4.1). Even though India is among the top eight nations in terms of gross R&D expenditure, in Purchasing Power Parity (PPP) terms it is behind all BRICS countries in terms of R&D intensity (R&D/GDP ratio). R&D intensity in India is marginally below South Africa but much lower than other BRICS nations; Brazil (1.2 per cent), China (2 per cent) and Russia (1.2 per cent) are ahead. Countries like South Korea (4.3 per cent), Denmark (3 per cent), Singapore (2.2 per cent) and the US are also far ahead (Basant 2021).

---

* These estimates are based on data from the Department of Science and Technology (DST), Government of India. DST estimates R&D expenditures on the basis of data collected from corporate firms (especially those that are recognized by the government for having R&D units), data from universities and public sector units and state outlays for R&D for different sectors through various state-owned labs and institutions.

Figure 4.1
National Expenditure on R&D in India (1995–2018)

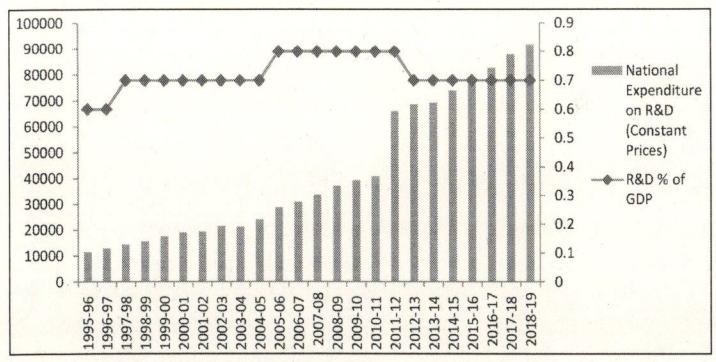

Source: Report on S&T indicators tables Research and Development Statistics 2019–20, Department of Science and Technology, Ministry of Science and Technology, Government of India.

Link: https://dst.gov.in/sites/default/files/S%26T%20Indicators%20Tables%2C%202019-20.pdf

Note: Figures from 1995–96 to 2010–11 are at base year 2004–05 and those from 2011–12 to 2018–19 are at 2011–12.

The other significant change in the post-reform period has been the increase in the share of the private sector in the nation's R&D. While the share of the public (central and state) sector declined from about 86 per cent to 56 per cent during 1990–2019, the share of the private sector increased from 14 per cent to 37 per cent (Figure 4.2). The share of the higher education sector had hovered around 4 per cent in recent years but has moved up to about 7 per cent in the most recent period. However, despite the increase in the share of the private sector, it is still much lower when compared to other countries. For example, the share of the private sector in China and South Korea has been more than 77 per cent in recent years. Similarly, the share of higher education in India is the lowest among the top eight R&D spenders in PPP terms. While early industrializers have a share that ranges between 12 per cent (Japan) and 26 per cent (UK), even China (7 per cent) and South Korea (9 per cent) are far ahead of India (4 per cent) in the comparable period (Basant 2021). As mentioned, only very recently, the share of higher education has reached close to 7 per cent. The decline in R&D intensity in recent years at the aggregate level is presumably because the slowing of the growth in *public* expenditure on R&D has not been compensated adequately by the increase in R&D expenditures by the *private* sector (Mani 2018).

Figure 4.2
National Expenditure on R&D by Sector (Percentage Share) (1990–2018)

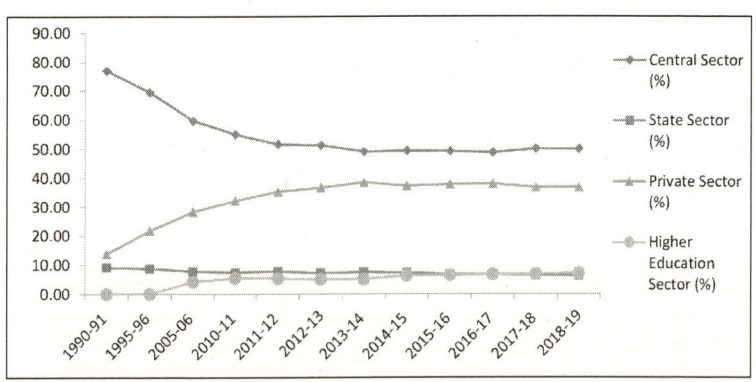

Source: Report on S&T indicators tables Research and Development Statistics 2019–20, Department of Science and Technology, Ministry of Science and Technology, Government of India.

Link: https://dst.gov.in/sites/default/files/S%26T%20Indicators%20Tables%2C%202019-20.pdf

Note: Higher education survey started since 2001–02 by DST.

These broad trends based on *aggregate* data observed in the post-reform period are also evident when one analyses the estimates of R&D intensity computed from *firm*-level data (Figure 4.3).* The increase is somewhat sharper in the 2000s as compared to the earlier period, although one observes a decline in the rate of growth post 2011–12. Overall, the increase has not been very rapid and R&D intensity remains low at less than 1 per cent of sales. In fact, there are only twenty-six Indian firms in the top 2500 R&D spenders in the world. Moreover, the distribution of R&D expenditures across firms has been quite skewed and only a few Indian firms spend heavily on R&D (Forbes 2017).

Disembodied technology purchases have also increased in the post-reform period. In absolute terms, according to RBI estimates, the expenditure on royalty and licence fees paid by India for foreign technology increased from $317 million in 2001 to $3719 million in 2014. The earnings from the sale of technology by Indian firms increased from $37 million to $525 million during the same period (Mani 2018). The disembodied technology trade balance seems to have widened in recent years. According to RBI estimates, the payments for foreign technology made by India increased from about $5 billion to $7.9 billion during the 2015–19 period while the receipts increased from $0.5 to $0.9 billion during the same period (CTIER 2020).

---

* It needs to be noted that the estimates used in this section are based on aggregates computed by the Centre for Monitoring Indian Economy (CMIE) from their firm-level database. Since the coverage of firms in the CMIE database has changed over time and in several instances reporting is sporadic and unclear, these should be considered as tentative. Despite the problems with the data, we expect the estimates to provide robust estimates of the *trends* in innovation inputs.

Figure 4.3

Sources of Knowledge (Inputs to Innovation) in Indian Manufacturing (1990–2018)

(Expenditure on Various Sources as a Percentage of Sales)

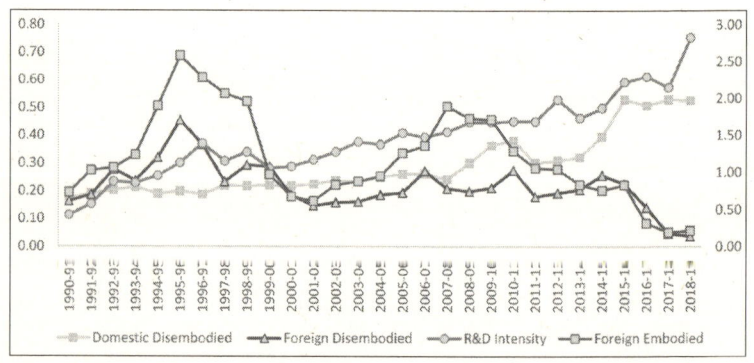

Source: Industry Outlook Database, CMIE.

Note: For all sources, except foreign embodied, the scale on the left is used.

Although the aggregate expenditure on foreign disembodied technology purchases through licensing has grown almost twenty-five-fold during 2001–19, its intensity of use shows a different pattern. Using CMIE aggregates of firm-level data, trends in R&D intensity are juxtaposed with the intensity of use of other sources of knowledge (Figure 4.3). A few interesting patterns can be observed during the 1992–2019 period: (a) Foreign technology purchases have been dominated by embodied technology purchase, i.e., import of capital. After a spurt immediately after 1991, the purchase of embodied *foreign* technology declined till about 2001–02, picked up a bit after that but started to decline again after 2008–09; (b) the purchase of *foreign* disembodied technology declined after the initial spurt in the early 1990s, picked up a bit after 2001–02 but declined sharply after 2014–15; and (c) reliance on licensing of technology from *domestic* sources has gone up consistently (albeit very slowly) after 2001 till 2014–15 but has increased sharply after that.

Overall, there is no clear trend in the intensity of use of different sources of knowledge or inputs to innovation after the economic reforms of 1991 which potentially enhanced both contagion and competition possibilities. At the aggregate level, the intensity of use of various inputs has fluctuated over time. The only exception is the use of domestic disembodied technology that has seen a consistent rise. One can argue that while R&D expenditures may be more uniform over time, technology purchases need not necessarily increase on a continuing basis, especially those related to embodied technology. The purchased technology may need to be absorbed and assimilated which would require firm-level efforts, some of which may get reflected in R&D expenditures. Spurts in technology purchases may be linked to spurts in the growth of the economy. There is some evidence to suggest that this was indeed the case at the aggregate level (Basant 2021). If we take the period 1992–2019 as a whole, R&D intensity in Indian manufacturing has grown faster than the intensity of use of external embodied and disembodied knowledge, both domestic and foreign (Table 4.1). This seems somewhat consistent with the line of argument above as R&D is required to absorb and adapt purchased technology.

Table 4.1

Trends in the Use of Various Sources of Knowledge by Industry Groups in the Indian Manufacturing Sector, 1990–91 to 2019–20

Intensity of Use (Expenditure to Sales) and Growth		R&D	Disembodied Foreign Technology Purchase	Disembodied Domestic Technology Purchase	Embodied Foreign Technology Purchase
Higher Than Average Intensity of Use	Higher Than Average Rate of Growth	1. Drugs and Pharma (3.28;0.55) 2. Electronics (1.36;0.37) 3. Transportation (0.91;0.43) 4. Beverages and Tobacco (0.43;4.03) 5. Construction Materials (0.39;1.36)	1. Transportation (0.61;0.55) 2. Electronics (0.37;0.58) 3. Machinery (0.27;0.38) 4. Diversified (0.27;1.84) 5. Consumer Goods (0.25;0.64) 6. Chemicals and Chemical Products (0.22;1.01)	1. Transportation (0.75;0.46) 2. Electronics (0.66;0.85) 3. Beverages and Tobacco (0.61;1.27) 4. Metal and Metal Products (0.45;0.49) 5. Consumer Goods (0.43;0.66) 6. Machinery (0.39;0.49) 7. Diversified (0.31;0.44)	1. Textiles (2.21;0.65) 2. Miscellaneous (2.20;0.73) 3. Electronics (1.96;0.62) 4. Metal and Metal Products (1.76;0.66) 5. Transportation (1.75;0.58) 6. Construction Materials (1.29;0.63)
	Lower Than Average Rate of Growth	1. Machinery (0.61;0.26)		1. Cement (0.98;0.33)	
Lower Than Average Intensity of Use		1. Diversified (0.30;1.13) 2. Consumer Goods (0.18;0.40)	1. Metal and Metal Products (0.16;0.78) 2. Miscellaneous (0.13;0.72)	1. Miscellaneous (0.14;0.51) 2. Chemicals and Chemical Products (0.10;0.57)	1. Chemicals and Chemical Products (1.15;0.71) 2. Diversified (1.12;1.15)

	Column 1	Column 2	Column 3	Column 4
Higher Than Average Rate of Growth	3. Miscellaneous (0.16;0.68) 4. Food and Agro-based Products (0.12;0.65) 5. Textiles (0.11;1.20) 6. Metal and Metal Products (0.11;0.49) 7. Cement (0.09;0.45) 8. Petroleum Products (0.08;0.50)	3. Petroleum Products (0.11;1.50) 4. Textiles (0.08;0.96) 5. Cement (0.08;0.80) 6. Food and Agro-based Products (0.07;0.87) 7. Drugs and Pharma (0.06;0.78) 8. Beverages and Tobacco (0.02;0.77)	3. Drugs and Pharma (0.07;0.63) 4. Textiles (0.06;0.62) 5. Petroleum Products (0.03;0.43)	3. Cement (0.88;0.98) 4. Consumer Goods (0.65;0.65) 5. Petroleum Products (0.63;1.16) 6. Beverages and Tobacco (0.52;0.57) 7. Food and Agro-based Products (0.33;0.69)
Lower Than Average Rate of Growth	1. Chemicals and Chemical Products (0.26;0.27)	1. Construction Materials (0.14;0.40)	1. Food and Agro-based Products (0.26;0.34) 2. Construction Materials (0.25;0.28)	1. Drugs and Pharma (1.05;0.53) 2. Machinery (0.81;0.46)
Average intensity of use for all Manufacturing	0.38	0.21	0.30	1.15
Average rate of growth for all Manufacturing	0.37	0.43	0.40	0.54

Source: Computed from Industry Outlook Database downloaded on 22/03/2021.

Note: 1. Figures in parentheses are estimates of the levels and growth rate in that order. Machinery includes electrical and non-electrical machinery.

There are, however, significant variations across industry groups both in terms of the intensity of use of various knowledge sources as well as in the growth in their use (Table 4.1). Not many industry groups with higher-than-average intensity of use of various knowledge sources have also experienced a higher-than-average growth in the intensity of usage. For the period as a whole, both the level of R&D intensity (in-house R&D expenditure as a proportion of sales) and its growth were higher than average in five industry groups—drugs and pharmaceuticals, electronics, transportation, beverages and tobacco and construction materials. While for the use of domestic disembodied technology, many more industry groups were in this category. Transportation, electronics machinery, consumer goods and chemicals showed higher reliance on foreign disembodied technology. Finally, use of foreign embodied technology was much higher than average and showed higher growth in textiles, electronics, metals and metal products, transportation and construction materials.

This evidence is consistent with earlier studies that showed that the pharmaceutical industry has become more research-intensive during the post-reform, post-TRIPS era by increasing R&D expenditures (Banerji and Suri 2017). However, although R&D expenditure has increased significantly in this period, this increase is not widespread across firms; in fact, more than 60 per cent of the Indian pharmaceutical companies do not spend at all on in-house R&D (Mishra 2011). At the same time, there is evidence to suggest that pharma firms are building manufacturing and other capabilities to enter foreign markets for drugs that are getting off-patent and they are coming up with a variety of other innovations that may not be captured by patents (Basant and Srinivasan 2016). So, certain sectors have indeed become more innovation-focused than others in the post-reform period.

The declining trend in the use of capital imports and also technology licensing (disembodied foreign technology purchase) in several industries is somewhat intriguing. With lowering of restrictions on imports and technology licensing, one expected a consistent increase in the use of foreign embodied as well as disembodied technology; easy access to foreign technology was likely to reduce the need for in-house R&D (Das 2004). Also, the broader expectation that economic liberalization would have a significant impact on various innovation activities does not seem to be getting fulfilled either. Without more detailed analysis, it is difficult to explain the inter-industry patterns in the sources of technology in the post-reform period. Bas and Paunov (2018), however, provide some interesting insights on the basis of detailed analysis of firm level data for early phases of liberalization in India. Their core argument is that liberalization did not affect all industry groups equally and therefore variations in firm responses across industries are to be expected. Their estimates suggest firms in liberalized industries were about 9 per cent more likely to invest in R&D than firms in non-liberalized industries. But, the impact varied with firm size; post economic reforms, firms in the top quartile were 23 per cent more likely to invest in R&D than those in the lowest size quartile. Trade liberalization also impacted R&D investments positively.

Moreover, in line with the *technology gap* argument discussed in the last chapter, more productive firms were likely to invest more in R&D in response to liberalization than less productive firms. Presumably, these firms perceived a higher possibility of 'catching up'. One of the reasons for firms not responding to economic reforms with innovation investments could be shortcomings in the business conditions (e.g., quality of public and private institutions) or location in less developed regions. Evidently, innovation activity of firms in response to the liberalization processes in India was more positive in states that were more developed. Besides, good business conditions were more important for smaller sized and less productive firms to undertake R&D in response to reforms than for larger sized and more productive firms (Bas and Paunov 2018). These findings suggest that the positive impact of liberalization on innovation investment is likely to be higher if the policy package also simultaneously improves access to better business services and business conditions. The importance of a good business environment is higher for innovation decisions of smaller firms and firms that are not close to the technology frontier, having to traverse a longer technology distance to compete.

Overall, firm-level data shows a faster growth in R&D expenditures than in foreign technology licensing. This seems inconsistent with the aggregate data which shows that the ratio of aggregate R&D expenditures and the amount spent on royalty and licensing fees for importing technology from abroad has declined significantly from about 13.6 in 2000 to about 2.2 in 2018 (Mani 2020). In fact, the increased royalty payments, especially by affiliates of MNCs operating in India, have caused some worry to the federal government, resulting in the formation of an inter-ministerial committee to look into this phenomenon.

Broadly, the use of various inputs or sources of knowledge can depend on a variety of factors. The evidence shared above does not indicate a very clear pattern. Given resource constraints, *at a point in time or for a short period of time*, enterprises may need to choose between *internal efforts* and *externally purchased inputs*. Over time, however, all mechanisms to acquire new knowledge are likely to be complementary, resulting in innovation-related outcomes. Therefore, the absence of any consistent trend in the use of various inputs to innovation may also mean that firms have been exploiting synergies across different sources of knowledge, using them at different points in time to respond to the changing market environment in the Indian economy since 1991. However, it is difficult to ascertain if this is indeed the case. In any case, the question why R&D expenses of the Indian corporate sector have not risen rapidly despite significant competitive pressure remains somewhat unresolved, although part of the answer could be that firms are combining different sources of knowledge and R&D may be riskier and more time-consuming than technology purchase. We shall revert to this issue later.

4.2.3. Output Measures

Patents are the most used output measures for innovation. Since the innovations reported in Chapter 2 were incremental and not new to the market, they are less likely to be outcomes of formal R&D and patented. Trends in patenting should be interpreted with that perspective.

Unlike input measures, which did not show a very consistent trend, patenting by inventors located in India has increased significantly in the post-reform period. The number of patents granted per year to Indian inventors by the US Patent and Trademark Office (USPTO) increased slowly and steadily during 1992–2019 and has seen a dramatic increase in recent years (Figure 4.4). In fact, the number of patents granted to Indian inventors by USPTO increased from a small number of twenty-four in 1992 to a fairly respectable number, 5378 in 2019. The number of US patents granted per year has exceeded 3000 since 2015. But India was still far behind China which had only forty-two US patents granted in 1992 but increased the number consistently to 19,209 patents in 2019.*

---

* The data has been compiled from the USPTO database. Link: https://www.uspto.gov/web/offices/ac/ido/oeip/taf/reports_stco. htm (downloaded on 14/07/2020).

Figure 4.4
Utility Patents Granted to India by US Patent and Trademark Office during 1992–2019

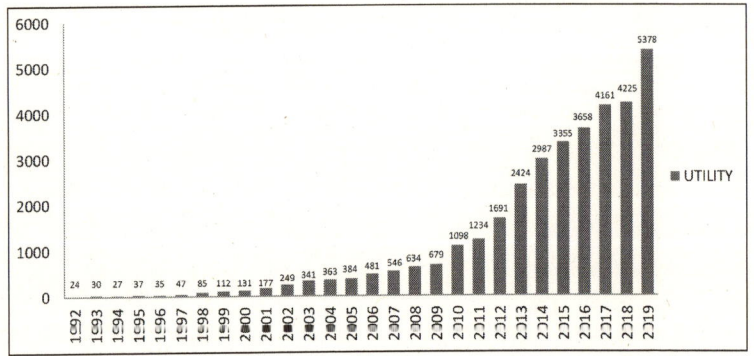

Source: USPTO database.

Link: https://www.uspto.gov/web/offices/ac/ido/oeip/taf/reports_stco.htm (Downloaded on 9/06/2020)

Since patent protection is nation specific, patent filings need to be made in all the countries where the inventor seeks protection. Available estimates suggest that patents filed from India in all jurisdictions increased from about 12,000 in 2009 to about 30,000 in 2018* (Figure 4.5). Since the same invention may have been filed in different jurisdictions, we cannot use these estimates as a measure of inventions in India. However, it is clear that inventions undertaken in India are being protected in multiple jurisdictions. For example, USPTO filings in 2009 were about 5000 while the total filings were about 12,000 but by 2018, the total filings had gone up to 30,000 while the

---

* Patents granted abroad increased from 1461 in 2009 to 6039 in 2018 for India but here again, the numbers were much higher for China for which the numbers increased from 3109 to 31,346 during the same period (CTIER, 2020).

US filings were only slightly more than 4000 (Figure 4.5).* Inventions undertaken in India are being increasingly protected in multiple jurisdictions suggesting that India as a technology base is participating actively in global inventive activity. However, a large share of patents granted by USPTO to Indian inventors can be attributed to the foreign R&D centres located in India; an estimate suggests that it can be as high as two-thirds of the total US patents of Indian origin (Basant and Mani 2012). Other studies have also shown that US MNCs have driven the recent growth of patenting from India (Chakrabarti and Bhaumik 2009; Mani 2020), spending by R&D centres set up as wholly owned subsidiaries (or partnerships) by MNCs in India has increased and many of the MNCs are now using India as a base to create their patents in new and emerging technologies.† (Dubey and Dubey 2010; Mani 2020).

Consequently, positive trends in the USPTO grants or trends in global filings may simply reflect that India has become an important location for R&D by MNCs and foreign patenting by domestic firms is still negligible and growing rather slowly.‡ Thus, it has been suggested the increased foreign patenting may not necessarily mean that firms in India have become significantly more innovative; rather it may simply become an important location for innovative activity

---

* There has been a concomitant increase in the filings for trademarks and industrial designs in various jurisdictions by Indian entities suggesting an overall increase in the use of multiple legal mechanisms to enhance appropriability (CTIER 2018).

† Apparently, a large share of the 9262 patents granted to IBM in 2019 was based on R&D conducted by the Indian affiliate of IBM.

‡ As mentioned, China is doing much better than India in US patenting. The sectoral distribution of patenting by the two countries is also different; while India focuses on pharmaceuticals and chemistry-related technologies, China has an important share of electronics and telecommunications (Basant and Mani 2012).

by external entities. While we will discuss the emerging role of FDI in R&D later in the chapter, it is useful to look at the trends in patents filed in India.

Figure 4.5
Patents Filed from India during 2009–2018 in the US and in All Jurisdictions

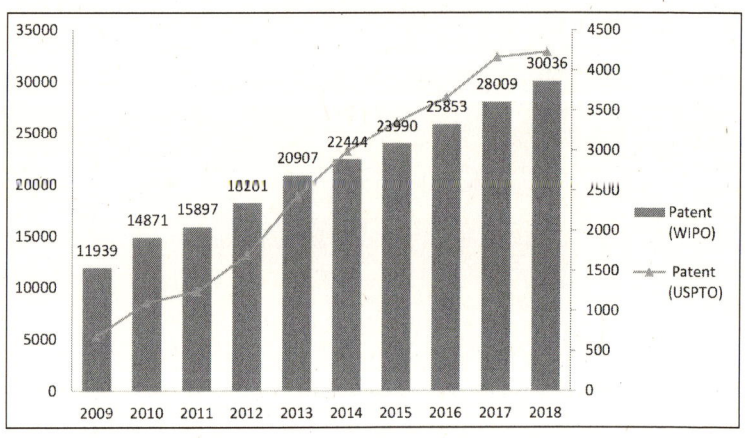

Source: World Intellectual Property Organization (WIPO) database and US Patent and Trademark Office (USPTO) database.

Link: https://www.wipo.int/ipstats/en/statistics/country_profile/profile.jsp?code=IN and https://www.uspto.gov/web/offices/ac/ido/oeip/taf/reports_stco.htm (Downloaded on 9/06/2020)

Note: For the WIPO database, the IP filings include filings by residents and non-residents in India and applications filed from India abroad (foreign filings). Since an application for the same invention can be filed in multiple jurisdictions, the WIPO estimates are overestimates as the same invention can be included more than once.

Patent filings in India have also seen an increasing trend. The number of applications increased from about 7000 in 1995–96 to almost 48,000 in 2017–18 (Figure 4.6).

Patent grants have also gone up but the number has fluctuated. The reasons for such large fluctuations are not clear but spikes in patent grants in 2007–08, 2008–09 and 2017–18 are particularly striking. One of the reasons could be the capacity to examine patent applications. The capacity seems to have been enhanced post 2010 as one observes a jump in the number of patent applications examined after that year. The increase in examinations has been particularly sharp since 2013–14. Interestingly, the share of domestic applications has risen quite rapidly in recent years (Figure 4.7). Immediately after the announcement of the impending change in the IP regime as India had decided to be TRIPS-compatible, the share of foreign applications rose in the 1990s and early 2000s.

Figure 4.6
Trends in Patenting Activity in India (1995–2017)

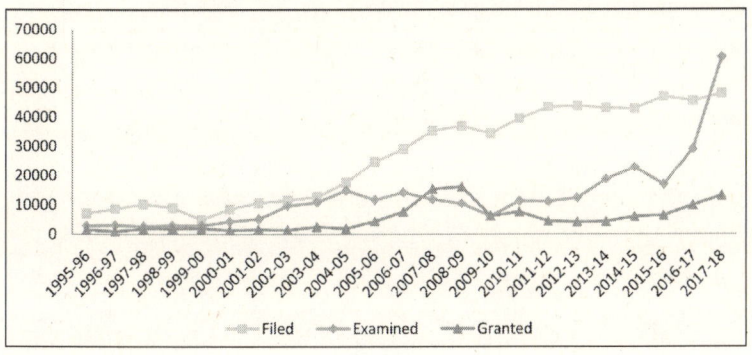

Source: Computed from Annual Reports of Intellectual Property India, the office of the Controller General of Patents, Designs, Trademarks and Geographical Indication, Government of India, downloaded on 15/06/2020.

Note: Figures prior to 2005–06 have been computed from Mueller, J.M. (2006), 'The Tiger Awakens: The Tumultuous Transformation of India's Patent System and the Rise of Indian Pharmaceutical Innovation', *University of Pittsburgh School of Law Review,* Vol. 68, pp. 491–641.

Figure 4.7

Patents Filed by Indians, Foreigners Resident Abroad and National Phase Application under PCT (1993–2017)

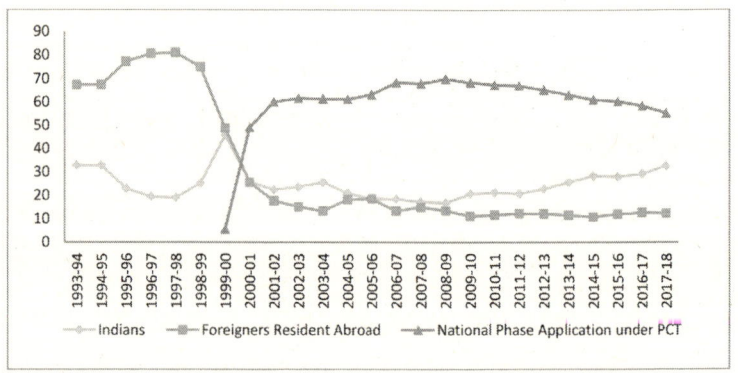

Source: Computed from Annual Reports of Intellectual Property India, the office of the Controller General of Patents, Designs, Trademarks and Geographical Indication, Government of India, downloaded on 15/06/2020.

Note: Figures for National Phase Application under PCT are given only since 1999–2000. Here PCT comprises both Indian and foreign residents abroad.

The areas of patenting in India have undergone a change in recent years (Annexure Table 4.1). In the years after the adoption of the new patent regime, one observes a rise in patent grants in the field of drugs and pharmaceuticals, their share in all patents granted being more than 20 per cent during 1999–2000 and 2003–04. But the share of patent grants in this field has declined in the subsequent period and was only about 6 per cent in 2017–18. While the share of patents in the field of mechanical and chemical inventions has remained reasonably high, the share of patents in the area of electrical inventions has declined. Patenting in the areas of computers and electronics and to a limited extent biotechnology is picking up.

As areas of patenting have proliferated and involve many new categories, the residual category of 'general' has seen a significant rise in its share. More detailed analysis suggests that within this residual category, the share of patent grants in the areas of communications, physics, biomedical and polymer science and technology has seen a rise (CTIER 2020).

Broadly then, while the aggregate expenditures on R&D and technology purchase have gone up, one does not observe a major change in the *intensity* of innovative activity among Indian enterprises in recent years as R&D and technology purchase expenditures as a percentage of sales have not shown any consistent trends. While the intensity of technology efforts has not grown at a rapid rate since liberalization began in 1991, the share of the private sector in overall R&D has gone up with a relative decline in the share of the public sector. There is a significant upturn in patenting activity but a large share of international patent applications is likely to be filed by R&D centres in India set up by MNCs. Significant improvements in innovative activities are restricted to a few sectors like drugs and pharmaceuticals, communications, transportation and electronics and IT. Both R&D and patent data suggest that Indian entities, private firms and public R&D institutions (mainly CSIR labs) are specializing in mechanical, pharmaceutical and chemicals industries. The focus on communications, computers and electronics is very recent.

At the same time, the bulk of patenting is done by MNCs and their focus area is the IT industry. Evidently, India is increasingly being used as a base for research in the IT industry by MNCs, enhancing the competition for research-related skills in the economy. Besides, the research areas of domestic and foreign firms do not overlap significantly. As mentioned, Indian involvement is much higher in the chemistry-oriented sectors than in information technology. The data also suggests that patenting activity did not match industry growth in telecom and IT till the early 2000s.[*] But that seems to have changed in recent years. Besides, there is also some evidence to suggest that MNCs are increasingly becoming active in the R&D space in biotech and pharmaceuticals through a variety of research-based alliances with Indian entities (Basant 2011). This raises the question of the capacity of domestic industry to benefit from spillovers generated by research activity undertaken by MNCs. The absence of such capabilities diminishes the likelihood that the domestic firm is *receiver-active* or ready to absorb and learn (Kodama, et al. 2007). This in turn would make the firms less likely to participate in innovation-related activities.

As such, the lack of a significant positive trend in R&D expenditures by Indian enterprises is a cause of worry and needs to be addressed. However, estimates discussed earlier suggest that the proportion of firms reporting innovation at the firm level is higher than the sharing of firms undertaking R&D. Evidently, many innovations are not based on internal R&D as was also discussed in an earlier chapter. It is also possible that they are a result of informal R&D that remains under-reported in the R&D statistics. There is a possibility, therefore, that the R&D estimates are underestimates and need to be investigated.

4.2.4. Technology Intensity of Exports from India

It has been discussed in the earlier chapters that high export orientation can result in innovation outcomes as firms from the exporting countries compete in demanding external markets.

---

[*] Athreye and Prevezer (2008) and Athreye and Puranam (2008) provide more detailed evidence for this. Also see Mani (2009).

Export-led growth of several East Asian economies and China are recent examples of this phenomenon. Since the reforms of 1991, India has done reasonably well in exports. Its share in merchandise (goods) exports has grown at 13.2 per cent per annum and the country's share in world exports has increased from 0.6 per cent in 1991 to 1.7 per cent in 2018. However, despite this growth, India was far behind China whose share was 12.8 per cent. Besides, merchandise exports as a percentage of GDP remained consistently lower for India compared to the world average throughout this period (Economic Survey 2019–20: p. 101).

While the export performance has been reasonable, the exports have been of low-end products. The shift in the technology intensity of India's manufacturing exports in the post-liberalization period has been rather slow. Technology-intensive industries include chemicals, pharmaceuticals, automotive and machinery, electronics, etc. According to some estimates, the share of high-technology-intensive exports in merchandise exports was less than 5 per cent in 1991 and has increased a bit to be in the range of 7–9 per cent in recent years (Mani 2017).* There is, however, some data to show that there is a significant shift away from low-technology-intensive exports to medium-low and medium-high-technology-intensive exports (Mann, Nayak and Aggarwal 2015). While the share of high-technology-intensive exports in merchandise exports has grown rather slowly and continues to be low, high-technology manufactures accounted for about 40 per cent of gross value added of the manufacturing sector

---

* High-technology merchandise exports from India mainly include automobiles, pharmaceuticals, electronics (especially telecommunications equipment) and aerospace products (Mani 2017).

in 2013–14 which implies that most of the high-technology products are targeted at the domestic market (Mani 2017). Domestic customers may not be as demanding as customers in export markets, especially those located in developed nations in the western world.

4.2.5. The Skill Deficit

Despite less than adequate growth in the innovative activities according to conventional measures, there are a couple of interesting developments. One, the Indian private sector is showing signs of becoming active in innovation-related activities, at least in some sectors, and the dominance of the public sector has declined in the process. Two, there is some evidence to show that the productivity of R&D investments in India is higher than in China, in terms of number of patents per unit of R&D expenditure (Mani 2009). While this result requires careful empirical analysis, it suggests that the human capital in R&D is of decent quality. But there have been several studies that underline the shortage of technically trained personnel. For example, Mani (2009) quotes industry reports to suggest that there is a shortage of about 25 per cent in skilled manpower for the engineering sector alone. According to some estimates, less than 5 per cent of India's workforce is formally skilled as compared to China where the share of the formally trained workforce is as high as 24 per cent. In some of the other developed nations this share ranges from 52 per cent in the US to 96 per cent in South Korea (Mallapur 2020).

On paper, India boasts of a large science and engineering manpower, but the average quality of the degree holders is not high. According to available estimates, in the year 2018 more than 8 million persons were awarded science and engineering (S&E)-related degrees and about 33 million were currently enrolled in such courses.[*] Given the indifferent quality of education, the major challenge has been the employability of the educated workforce. The India Skills Report, 2019 estimates that on average, only about 48 per cent of the educated workforce is employable; the employability ranging between 57 per cent for engineers and 18 per cent for polytechnic graduates.[†]

One area where the skill gaps might hurt India the most in the immediate future is that of ICT as it is becoming quite ubiquitous with innovations in digital technologies transforming virtually all activities that one undertakes. There is a shortage of educated workforce with specific technical and soft skills that are now required in this sector. According to some estimates, between 60 and 70 per cent of the current ICT workforce in India would require retraining due to the changing skill requirements in the market. Apart from the rapidly changing technology, which results in high obsolescence rates for the technical components of the academic curricula, poor instruction in mathematics and absence of interdisciplinary education or curricula have been identified as some of the key reasons for the skill gap or shortage (ILO 2019). The relevance of multidisciplinarity for innovation will be discussed later but one mechanism to improve innovation capability, especially the ability to absorb, can be skill training by the corporates themselves

---

[*] Of the 8 million graduating persons, more than 1.5 million had postgraduate or PhD degrees and of 33 million currently enrolled, more than 4 million were enrolled in postgraduate (including PhD) courses (CTIER 2020).

[†] The employability for MBAs (36 per cent), general graduates (35 per cent) and postgraduates (43 per cent) was in between (see India Skills Report, 2019, Change this to: https://www.aicte-india.org/sites/default/files/India%20Skill%20Report-2019.pdf).

if educational institutions are not providing training that is appropriate for the emerging needs. But even the intensity of skill training undertaken by enterprises has not seen any significant upturn in recent years; the percentage of sales invested in training has typically not been higher than 0.2 per cent (Figure 4.8).

Figure 4.8
Trends in Skill Training Expenditures (1992–2018)

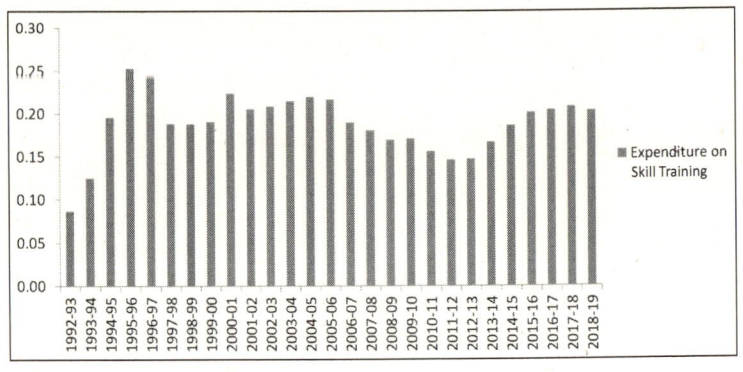

Source: Computed from Prowess Database downloaded on 15/06/2020.

Note: Figures are in percentage (Skill training/Sales of goods*100) and three-year moving average method had been used to compute the data.

A variety of policy initiatives to build vocational skills have been tried over the years without much success. The most recent Skill India programme* does not seem to be an exception. A recent report (Government of India 2016) has pointed out numerous shortcomings in India's vocational education and

---

* For details of this policy initiative, see https://skillindia. nsdcindia.org/

training systems. These include many issues that have been flagged for several years. Multiple government agencies offer vocational training through a plethora of organizations with insufficient financing, limited training capacity and infrastructure, and without nationwide Vocational Education and Training (VET) standards.* Moreover, there is a lack of an integrated on-site apprenticeship training which is essentially due to inadequate industry interface. According to this report, Sector Skill Councils (SSCs) that were mandated to assess skill needs of different sectors and create appropriate curricula and training programmes to fulfil these needs have failed to do so. The SSCs are expected to provide the interface between employers, trade unions and governments, and in the process, in the process, help vocational education become meaningful for industry. Industry associations have been driving the SSCs but that has not resulted in adequate involvement of the employers.

This outcome is quite surprising as the key organization for skill development, the National Skill Development Corporation (NSDC) was created as a unique public-private partnership for mobilizing resources from industry, financial institutions, banks, multilateral and bilateral external aid agencies, private equity providers and different ministries and departments of central and state governments. NSDC is financed by the National Skill Development Fund (NSDF), which was set up by the Government of India for raising funds from government and non-government entities for skill development in the country. Evidently, despite being in

---

* Evidently, there are seventeen ministries, in addition to the Ministry of Skill Development and Entrepreneurship, which undertake skill development. But none of them owns the National VET standards (Government of India 2016).

the PPP mode, the private sector has not contributed to the work of NSDC. In fact, more than 99 per cent of the NSDC funding was provided by the Government of India which means that they have not been able to mobilize resources from the multiple sources mentioned above (Government of India 2016). The lack of private sector participation implies that they do not see value in this endeavour, probably because despite being in the PPP mode, its functioning is bureaucratic. As a consequence, the involvement of user industries in identifying training needs and developing programmes seems to be very limited. Apparently, state resources have been used to support private entities for short-term training which does not have much market value as industry participation has been rather poor.* It has been suggested that the SSCs should be owned, funded and driven by the sectoral employers and not by industry associations to facilitate more intense industry participation. Such a model is also essential for operating SSCs in a truly PPP mode.

As the current PPP model of NSDF-NSDC does not seem to be working well, it is critical to think about an appropriate organizational model to create and sustain a PPP in the vocational training space. Studies have pointed out several organizational lacunae in the current model, including issues related to conflict of interest between the operations of NSDF and NSDC as well as lack of accountability of NSDC (Government of India 2016). Once an incentive-compatible organizational structure is in place, the private sector may also contribute to help set up a large number of VET institutions. Many of these can be run by the private sector

---

* Government of India (2016) provides a detailed critique of the vocational training system in India.

to provide long-term competency-based courses according to the emerging requirements of industry.

The skill gaps not only highlight the increasing role of VET institutions but also of higher education as a key instrument of innovation policy as the Indian economy gets integrated with the global economy and the competition for the skilled persons to enhance innovation-related activities becomes more intense. The ability of Indian firms to undertake and/ or participate in global production and technology networks would also largely depend on the availability of adequately skilled manpower.

4.2.6. Foreign Direct Investment as a Source of Innovation

As discussed in the earlier chapters, FDI can influence innovation in a host economy through competition and contagion (spillover) effects and the type of FDI would determine the nature of this effect. For example, sectoral distribution, technology intensity, mode of entry (M&A vs greenfield), nature of activity (assembly vs local manufacturing) as well as type of domestic linkages of MNCs would affect the flows of knowledge their presence might entail. It has long been recognized that FDI can potentially be a major source of technology for developing economies. FDI, whether in a wholly owned subsidiary or in a joint venture with minority/majority local participation, is often seen as the only way to obtain the latest technology-related information from abroad. However, it has been argued that while such MNC participation can provide a transfer of technological information to a developing country, it may not build technological capabilities to understand this information well. It has also been recognized that the lack of local control can potentially have many deleterious outcomes for local technological development and such development is unlikely in the absence of local efforts (Dahlman, Ross-Larson and Westphal 1987).

Some evidence for other nations, discussed in the last chapter, showed that a liberal FDI policy combined with a stringent IP regime increased the technology transferred to MNC affiliates, R&D expenditures of affiliates as well as the level of foreign patent applications. Given the policy changes in India, this would suggest MNC activity here should result in better technology transfer and create larger potential of knowledge spillovers. But there are very few studies that explore the impact of FDI on innovation activities in India. Kathuria (2008), examining the impact of FDI on R&D investments by medium and high-tech firms in India, found that FDI flows had a negative impact, at least initially, on R&D. He argued that this was the case as economic reforms allowed such firms to import technology instead of investing in development of innovation. The analysis of the trends in imports of technology and local R&D undertaken earlier in the chapter did not show a very clear pattern but that does not inform us about the longer-term impact of FDI on the R&D efforts of Indian firms. This would require a very detailed exercise that has not been undertaken here. Instead, a few characteristics of FDI in India in recent years that might have significant implications for innovation-related activities are highlighted.

The FDI inflows into India have risen dramatically since the economic reforms of 1991; the rise in inflows has been particularly significant since 2005–06. Average annual FDI increased from $1.7 billion (1991–2000) to $2.9 billion (2000–05) and then dramatically $57 billion per year during 2014–19 (Figure 4.9). Interestingly, disinvestment (repatriation of capital) has also been on the rise in recent years.

In fact, from about 15 per cent in 2009–10, equity disinvestment as a proportion of equity inflows rose to about 47 per cent in 2017–18. Consequently, net flows have been much lower than the gross inflows reported each year (Rao and Dhar 2018). Moreover, share of greenfield investments in FDI showed a declining trend after 2000 till about 2013 and M&As were the preferred mode of entry by MNCs during this period* (ISID 2014). There seems to be some movement towards greenfield projects since 2014 and about 40 per cent of FDI came through this route during the last six years (Anand 2020). While the share of FDI in manufacturing increased in the 2000s, a large part of these inflows was through M&A. The top three manufacturing industries that received FDI were drugs and pharmaceuticals, chemicals (other than fertilizers) and the automotive sector. Moreover, the share of FDI in high-tech manufacturing sectors was only about 27 per cent of total FDI in manufacturing during 2003–14 and more than 80 per cent has come through the M&A route (ISID 2014). The situation does not seem to have changed in recent years (Rao and Dhar 2018). Finally, even within manufacturing, often the focus seems to be on assembly of products for sale in the domestic markets with little interest in exports. In more recent years (since 2014), the share of services has been on the rise (Rao and Dhar 2018). Finally, very little is known about the linkages the MNCs have built post economic reforms and the information on their contribution to training is also not easily available.

---

* Some studies have even suggested that the share of FDI through M&A mode is underestimated for this period. For example, officially, during the 2003–13 period, M&As accounted for less than 30 per cent of the inflows in manufacturing. But in reality the share is likely to be in the range of 47–54 per cent (ISID 2014).

Figure 4.9
Foreign Direct Investment (FDI) Inflows (Equity Capital Only) in India (1991–2019)

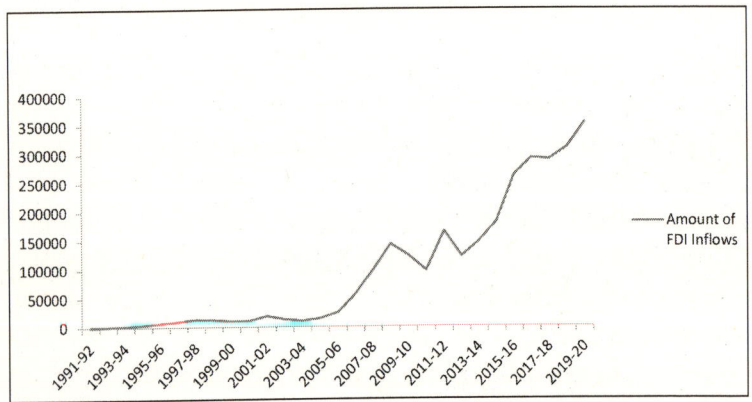

Source: Indiastat Database downloaded on 17/06/2020.

Note: Figures are given in INR crore.

Given this lack of information, it is difficult to ascertain if knowledge flows through FDI have facilitated capability-building among domestic firms. For instance, a large proportion of exports of manufactures from China is contributed by MNCs operating from China. This is unlikely to be the case in India although there is now some evidence to show that MNCs' share of export production is increasing although concentrated in specific industries such as the automotive industry, for instance (Basant and Mani 2012).

Firm-level data on the use of various sources of knowledge when aggregated by firm ownership provides some interesting patterns (Table 4.2). On average, the R&D intensity of domestic firms in India (both public and private) has been somewhat higher than that of foreign firms during the post-reform period. However, the intensity of use of domestic disembodied technology through licensing is somewhat higher for foreign private firms than for private domestic firms; the public sector firms have used this knowledge source most intensively. The intensity of use of foreign disembodied technology is more or less the same across all categories of firms, it being marginally higher for domestic firms, especially public sector firms. Use of disembodied foreign technology through capital imports is much higher for domestic private firms than foreign firms. While these estimates may be tentative and at times, the differences are not significant in statistical terms, foreign firms do not seem to be using foreign sources of knowledge (both embodied and disembodied) more intensively than domestic firms. In fact, the converse seems to be the case. Moreover, foreign firms seem to be using domestic knowledge sources more intensively than the domestic firms.

Table 4.2

Sources of Knowledge in the Indian Manufacturing Sector by Ownership Categories, 1992–93 to 2018–19

Sources of Knowledge	Average Expenditure as Percentage of Sales			
	Private Domestic	Public	Private Foreign	All Firms
In-house R&D	1.1 (0.4)	1.1 (0.5)	0.9 (0.6)	0.9 (0.3)
Disembodied Domestic Technology Purchase	1.6 (0.2)	3.9 (0.4)	1.9 (0.2)	1.8 (0.1)
Disembodied Foreign Technology Purchase (FOREX Spending as Royalty and Fees)	1.4 (0.4)	1.6 (1.1)	1.3 (0.3)	1.2 (0.2)
Embodied Foreign Technology Purchase (FOREX Spending on Capital Imports)	3.4 (0.3)	2.6 (0.6)	2.5 (0.3)	3.0 (0.3)

Notes: 1. Figures in parentheses report coefficient of variation.

2. Private domestic includes business group-affiliated firms.

3. Public firms include both Centre and state-owned entities.

Source: CMIE, Prowess (based on data collected on 04/04/2020). Computed by the author from firm-level data. The estimates for all firms do not match with the estimates reported in Table 4.1 which are based on firm-level aggregates provided by CMIE. The sample of firms used by CMIE and by the author is likely to be different. While the absolute estimates differ, the broad pattern across category of firms is likely to be robust.

While it is difficult to interpret these aggregate estimates to tease out technology strategies of different categories of firms, on average, excessive reliance of foreign firms located in India on foreign embodied and disembodied technology is not evident. In fact, their relatively higher use of domestic disembodied technology suggests that they are building technology linkages with local players and find domestic technological capabilities useful. From the perspective of policy, this feature can be seen in two ways. One, foreign firms building technology linkages with local firms would result in knowledge flows and associated contagion (spillover) effects. Two, the strategy of foreign firms not bringing in foreign technology may reduce the contagion effect (spillover potential) of their presence and even reduce the competition effect! But overall, these patterns seem to be consistent with the fact that the bulk of recent FDI activity has been through M&A, and MNCs may not have brought in new technologies through foreign technology licensing and capital imports.

4.2.7. India as a Host of MNC R&D

Apart from the rapid increase in FDI inflows into India in recent years, at least to some extent, the nature of MNC involvement in India seems to have moved up the value chain. The overall evidence is somewhat mixed; one MNC activity is clearly high up in the value chain and that is undertaking R&D in India. The number of foreign R&D centres in India has shown significant growth in the post-reform period. A majority of the R&D centres are either subsidiaries or branches of US-based MNCs, and are quite active in the ICT sector (Basant and Mani 2012). According to some estimates, R&D investment by US MNC affiliates was of the order of $18 billion during 2012–2017.* The bulk of this R&D must

---

* This estimate is based on the data available at the website of the Bureau of Economic Analysis, Survey of US Direct Investment Abroad (Annual Series).

have taken place in these MNC R&D centres. There are more than 1000 MNC R&D centres in India today (Nabar 2018). These represent more than 80 per cent of the top 100 global R&D spenders and about 50 per cent of the top 500 global R&D spenders (Forbes 2017). A recent study has estimated MNC R&D expenditure in India to be in the range of $7.8–8.4 billion, which is quite high and effectively more than double the R&D efforts by industry in India (Nabar 2018). What drives such FDI and how does this affect building of innovation capabilities in the country?

A large variety of factors have been identified by earlier studies that might influence the incidence and extent of R&D undertaken by foreign firms in developing countries. These factors can be divided into two broad categories (Basant and Mani 2012):

i. *Pull* factors, essentially driven by *demand* in host (developing) countries, resulting in the use of *market-seeking* or *knowledge-exploiting* strategies of MNCs. Firms use/adapt their existing knowledge to satisfy demand in host nations; and

ii. *Push* factors, driven by lack of adequate *supply* of knowledge and other resources in home countries while such resources are available at competitive rates in host nations. MNCs employ *resource-seeking* or *knowledge-augmenting* strategies as they undertake R&D in developing nations. These strategies may not be used to cater to host country markets.

A survey of MNC R&D centres in India undertaken a few years ago showed that while these centres fulfil a multiplicity of objectives and undertake a wide range of research activities (including high-end work in basic research and product design), the MNCs' primary objective is not *market seeking*; the *resource-seeking* dimension seems to dominate. Development of new technologies for global and regional markets is more important for these centres than modifying or adapting technologies for local market needs or manufacturing requirements. Local researchers provided significantly in terms of research ideas to be worked upon in the centres which are exported back and patented in the home countries. Availability of quality scientists and engineers at low wages as compared to their home countries is one of the important determinants of their location in India (Basant and Mani 2012). Recent developments, however, seem to suggest that location of MNC R&D centres in India is not only used for its low cost of operations but also for developing technologies for markets like India (Nabar 2018). It is not yet known if this shift in the market orientation of the research undertaken by the MNC R&D centres has also resulted in changes in certain other features that were observed earlier (Basant and Mani 2012):

i.   While the centres did not perform low-end operations and many research outputs are patented in host and home countries, the projects performed in these centres are small, and have short-term horizons of less than two years. It is possible, however, that most of the R&D centres are primarily performing the more labour-intensive parts of a large R&D project with only a few implementing the entire R&D project; and

ii. The linkages of these R&D centres with local enterprises and institutions (including HEIs) are rather limited. Both for performing R&D and for solving research problems, they seek significantly more support from the global business units of the MNC. So knowledge spillovers for the local economy emanating out of the activities of these centres may not be non-existent but remain rather limited due to the limited interaction with local entities.[*]

The importance of India as an R&D hub is also evident from the data on the export of R&D services. Such exports have been increasing by around 40 per cent in nominal terms per annum during the period 2004-18: exports of R&D services which were only $118 million in 2004 increased to about $4 billion by 2018 (Mani 2020).

A useful policy question is if these centres create spillover (contagion) benefits and build capabilities through various linkages. No information is available on the circulation of R&D personnel across enterprises. India does not have any explicit policies to promote FDI in R&D although R&D tax incentives are available to both foreign subsidiaries and domestic firms. The utilization of such investments by MNCs, however, is not known. Stand-alone MNC R&D centres presumably are not eligible for such subsidies. However, R&D services *excluding* basic research and setting up of R&D or academic institutions that award degrees/diplomas/certificates are allowed 100 per cent FDI under the automatic route. This must have eased the regulatory burden of setting up such R&D centres in India by MNCs. IP policy seems important as many technologies that are being developed are patentable.[†]

---

[*] Available studies on China (Lan and Liang 2006) have also shown that foreign R&D centres are hardly connected with the national system of innovation of China as their linkages are often enough with their own parent firms abroad.

[†] The empirical results of He (2007) also showed that at the cross-country level, stronger IPR regimes and good contract enforcement has a positive impact on decisions to set up R&D centres in developing countries.

It is very difficult to analyse the implications of the patterns of MNC activity enumerated above, for contagion-competition effects. Till recently, MNCs have not entered high-tech areas in any significant manner, nor have they been very active in creating state-of-the-art greenfield projects. The opportunities to learn through contagion effects thus seem somewhat limited. They have, however, created significant competition effects. The ability of domestic firms to respond to this competition would determine if innovation can be an outcome of the increasing competitive pressures.

The emerging phenomenon of setting up R&D centres in India is quite interesting and can create learning opportunities. Based on the type of research activity undertaken by such centres, these have been categorized as *support labs* (offshoring of R&D by the parent company), *locally integrated labs* that involve R&D exports and local manufacturing and marketing activities, *collaborative labs* that collaborate with local entities and *internationally independent labs* whose research agenda is not driven by the parent companies (Krishna and Bhattacharya 2009). Earlier surveys suggested that the R&D centres in India were essentially 'support labs' catering to the needs of the parent company and their 'local integration' through various linkages and 'collaborative activities' was limited. In fact, there is little or no evidence of these MNC R&D centres interacting with Indian firms and institutions (Mani 2020).

While the existence of such labs can enhance the competition for research personnel between domestic firms and MNCs, the knowledge spillovers would essentially depend on the nature of local linkages and collaborations. Policy in India seems to enable, at least to some extent, the setting up of such centres through the development of a reasonably skilled research workforce and easing entry conditions. However, it is not entirely clear how the linkages between the R&D centres and the local entities can be built. Policy initiatives to make the HEIs more research-oriented and vibrant may help. We shall revert to this issue later.

4.2.8. Emerging Role of Information and Communication Technologies (ICTs)

ICT has seen significant digital innovations in recent years. As discussed in the last chapter, adoption of ICT along with various digital technology-related innovations has affected the way enterprises undertake their business and organize their innovation-related activities. Studies have highlighted the positive effects of ICT adoption on productivity (Erumban and Das 2016; Commander et al. 2011). Adoption of ICT, however, has been affected by a variety of factors, including, availability of telecom and power infrastructure and access to capital (Basant et al. 2006; Basant 2007; Houben and Kakes 2002). Consequently, ICT adoption has differed a great deal across nations and regions within a nation. Given the fact that ICT itself is a technological innovation, its adoption can be seen as innovation at the firm level. Moreover, ICT and many digital technologies which work through ICT infrastructure also affect innovation as they support the creation of new and better applications, production processes and organizational practices. With the possibility of using ICT for various functions within an enterprise—product/service development, manufacturing, marketing, distribution as well as after-sales support—it can enhance the productivity of all inputs. Therefore, adoption of ICT can play a significant role in innovation by enterprises.

According to a survey, the use of computers in Indian enterprises is not very widespread. Only in about 5 per cent enterprises, more than 75 per cent employees use computers regularly. Less than 20 per cent employees use computers regularly in more than 52 per cent enterprises (Table 4.3). But this may be an inappropriate measure of ICT usage among Indian enterprises as a large proportion of production workers may not be required to use computers. Use of ICT for different functions provides a better picture of ICT adoption (Table 4.4). Interestingly, more than 96 per cent enterprises use the Internet to connect with their customers and suppliers but the use of the Internet for internal communication among enterprises was restricted to about 68 per cent firms. What is remarkable is that a significant proportion of enterprises (56–73 per cent) use ICT for marketing and sales, inventory management and even R&D. More than 50 per cent enterprises have software customized for their specific needs, either purchased or developed in-house. Interestingly, some recent studies suggest that the rate of adoption of AI technologies and the extent of their use by Indian firms is better than many developed economies (Nabar 2018).

Table 4.3
Distribution of Enterprises by the Percentage of Employees in the Establishment Regularly Using Computers, 2013

Percentage of Employees Using Computers	Percentage of Enterprises
None	4.0
Less than 10 per cent	38.6
10–20 per cent	9.5
20–50 per cent	44.7
50–75 per cent	5.9
75–100 per cent	5.0

Source: Computed from World Bank Enterprise Survey—India (Innovation Module).
Sample size=3492.

Table 4.4
ICT Usage in Indian Manufacturing and Services Firms, 2013

Purpose	Percentage of Enterprises
Purchase or in-house development of customized software	51.2
Use of Internet for e-mail with clients and suppliers	96.5
Use of Internet for internal communication among employees	67.6
Online purchase of inputs or finished goods purchased to resell	55.9
Online sales and order fulfillment	58.9
Manage inventory	72.8
Marketing of products and services	65.4
R&D for ideas on new products and services	72.7

Source: Computed from World Bank Enterprise Survey—India (Innovation Module).
Sample size=3492.

The use of ICT can vary in terms of the *extent* (diversity or spread of functions for which it is used as depicted in Table 4.4) as well as the *intensity* of use of various functions. To capture this empirically is a difficult task. Basant et al. (2006) try to measure it by differentiating between: (i) non-use of ICT; (ii) use of ICT for only some office functions along with accessing the Internet and e-mailing; (iii) use of ICT for some advanced applications and where most processes are automated but there is no integration into a central system; (iv) most processes are automated and some of them are integrated into a central system; and (v) almost all processes are automated and integrated with a central system. They found that firms vary a great deal by intensity of use based on this five-fold categorization. More interestingly, it was found that the productivity benefits to the enterprises accrued only after a certain threshold of ICT adoption intensity was crossed. In terms of the categorization above, productivity gains emanating from ICT adoption among Indian enterprises kicked in only after level three was crossed. The extent and intensity of ICT adoption can also be viewed as changes in organizational and other practices as with the adoption of ICT, the ways of doing things change. In that sense, changes in ICT can also be seen as a *practice* innovation that facilitates *product* and *process* innovations and also adds to their efficacy in terms of benefits. Seen from this perspective, ICT is an important part today of the innovation ecosystem as well as the innovation process, and its adoption can be critical for innovation development and diffusion.

Several studies have highlighted the role of ICT in disseminating knowledge about innovations. There is also evidence to show that a variety of *practice*-related innovations may not have been introduced by Indian firms due to constraints on adoption of ICTs. A comparison of ICT adoption in Brazil and India has shown that weak institutions and infrastructure together result in lower adoption and lower returns to ICT adoption, particularly in India (Basant 2007). The regional variations in Brazil were found to be far smaller. However, firms in India located in states with better institutions and infrastructure had returns to ICT that were close to those obtained by Brazilian firms. These results are consistent with those of the studies discussed earlier that R&D investments, especially for smaller firms, are affected by business and institutional conditions. All this suggests that much of the policy challenge in India consists of addressing the sources of these inefficiencies and institutional weaknesses at the state level, both for ICT adoption and for supporting innovation-related activities.

Insights from some case studies of ICT adoption in Indian SMEs show that initial training costs of firms rise with ICT adoption (Basant 2006). Since required skills are not so easily available in the market, firms have to internalize training costs. The problems associated with power disruptions create significant constraints on ICT adoption.* Moreover, despite significant reduction in hardware and software prices, many firms (especially SMEs) find the costs of ICT adoption to be high, especially when they plan extensive usage and a high intensity of use. While continuing reduction in tariff rates for hardware and software would certainly help, the key to reducing prices, however, is to rapidly increase the market base. That might require some incentives for ICT adoption

---

* As in the case of training, power generation is also being increasingly internalized by several firms in India. Given the high cost of in-house training (and the associated externalities) and power generation, not all firms are able to afford it and ICT adoption and the related productivity gains suffer in the process (Basant 2006).

but will need to be thought through. As more and more enterprises, especially SMEs, adopt ICT, prices may come down further as such adopters are much more cost-conscious and often require modular and customized solutions. And as a consequence, one might see significant adoption of *practice* innovations and a breeding ground for *product* and *process* innovations.

Unfortunately, there has been no detailed information on the trends in the extent and intensity of use of ICT by Indian firms. A recent study (Krishna et al. 2018), however, provides some aggregate estimates to suggest that there has been a significant increase in the total ICT investment (hardware, software and communication) in the Indian economy over the years. ICT investment increased from about INR 7995 crore in 1990–91 to INR 40,475 crore in 1999–2000, a five-fold increase in a decade. The increase in the subsequent period was even sharper with investment reaching INR 2,84,874 crore in 2012–13. Despite this rapid increase, ICT investment as a proportion of Gross Fixed Capital Formation (GFCF) in the economy has not increased significantly over the years. In fact, the share of ICT investment in GFCF has been quite volatile, ranging from 5.7 per cent in 1990–91 to 11.6 in 2008–09. After reaching the peak in 2008–09, it has seen a continuous decline.

Absolute level of nominal investment in ICT in the organized manufacturing sector has also increased over time. It increased continuously from about INR 650 crore in 1998–99 to INR 3122 crore in 2008–09. After a decline in 2009–10, it continued to increase again, reaching INR 4144 crore in 2013–14 (Krishna et al. 2018). As was the case with aggregate ICT investment, investment in organized manufacturing as a share of GFCF in the sector has also been quite volatile; after reaching a peak of about 2.3 per cent in

2003–04, the share has shown a declining trend till 2014. More recent data is not readily available. Interestingly, the composition of ICT investment in the manufacturing sector has changed over the years. Chemicals, textiles and food products industries contributed the bulk of ICT investment in 1993–94 with a share of 49 per cent. By 2013–14, new industries emerged as the dominant ICT investors with machinery, transport, electrical and optical equipment having a share of about 48 per cent of the total ICT investment in India's organized manufacturing sector (Krishna et al. 2018).

Overall, while there has been a significant increase in ICT investment in absolute terms during the last two decades or so, the relative importance of aggregate ICT investment in total capital formation in India has not seen a consistent and significant increase. The same trends are being observed for ICT investment in the manufacturing sector. In fact, both in the aggregate and in the manufacturing sector, the share of ICT investment in GFCF has seen a decline in recent years. Insofar as ICT investments are important for adopting and developing innovations, constraints to ICT adoption need to be identified and removed. Apart from the telecom and power infrastructure, which is critical for ICT adoption, policy may need to focus on hardware and software prices along with the availability of skills. Some studies have suggested that the relatively low levels of ICT adoption in India can be explained to a significant extent by the absence of relevant skills (Erumban and Das 2016). Universities are a major source of building these skills.

4.3. University–Industry Linkages (UILs) in India

In the discussion above, the role of well-trained persons in innovation-related activities has been highlighted, including for activities relating to ICT adoption. Provision of skills is one of the primary objectives of educational institutions. But as the discussion on NIS and UILs in Chapters 2 and 3 suggests, the role of universities or HEIs in innovation is much broader than provision of skills and links between higher education and innovation take different forms. To recapitulate, skills and training build capabilities and policies that foster such capabilities (e.g., education and S&T policies), and can intensify competition by generating knowledge for innovation or technology-based entry, and facilitate contagion by building capabilities to absorb technology and exploit spillover potential. Existence of technological capabilities (e.g., trained S&T personnel) can attract innovation-intensive investment and facilitates building of linkages between innovators and others, thereby enhancing competition as well as contagion potential. Thus, policies relating to education and science and technology, which impinge on innovation capabilities and provide incentives for commercialization of technologies developed in HEIs and building UILs, affect all three stages of technological change—invention, innovation and diffusion.

At a broader level, HEIs can potentially provide a common and possibly neutral platform for discussion about the broader goals of innovation policy and a forum where there can be relatively open interaction between industry and government. This may be quite valuable in developing countries where the relationship between the bureaucracy and industry is often either antagonistic or clientelist, both of which preclude a productive dialogue. This function may be especially important in situations where the country is moving from one type of industry-government relationship to another, e.g., during the process of liberalization (Basant and Mukhopadhyay 2010). This role is rarely played by

HEIs in India, partly because of the lack of autonomy and limited credibility that is getting continuously eroded by the politicization of decisions with respect to higher education. The lack of autonomy and erosion has been documented by several studies (e.g., Kapur and Mehta 2017; Chandra 2017) and we shall revert to this issue later.

Four types of linkages were identified earlier through which HEIs can contribute to the innovation process: (i) supply of skilled persons; (ii) creation of knowledge; (iii) creation of new enterprises; and (iv) supply of services that contribute to innovation. While we will discuss the enterprise creation part of the linkage in some detail in the next question, issues relating to the other three links are being addressed below.

Available evidence from India and other developing countries suggests that labour market linkages (including competency development and training) between industry and academia is the most prominent link as universities contribute relatively little to patenting, licensing and new enterprise creation, except to a very limited extent in life sciences (Basant and Mukhopadhay 2010; Yusuf and Nabeshima 2007; Basant and Chandra 2007b; Krishna 2012; Krishnan 2012). Given the relatively poor state of educational infrastructure in India, the predominance of state funding for higher education which is essentially used for salaries and low levels of funding of R&D by industry, creation of knowledge has not been a strong channel either.

Supply of services for innovation-related activities (testing, training, prototype development) is especially important for developing countries like India (where the size of the firms is typically small and the numbers relatively large), for two reasons. First, for a small firm, developing such skills in-house is more difficult compared to a large firm. Second, the talent pool is relatively limited and unable to cater to the large number of firms. HEIs such as universities can potentially provide these services as a common pool resource, aggregating the limited talent and making it available to all firms on a fee-for-service basis. This role has largely remained restricted in India to some short-term training (Basant and Chandra 2007; Krishnan 2012) and corporates often complain that they do not get the necessary skills that they require and often have to undertake training on their own (Reddy, Xie and Tang 2016).

Limited research focus and scant research output of HEIs is at the core of their inability to play a critical role in the innovation ecosystem of a developing country like India. One argument has been that if regulation permits HEIs to own intellectual property on the outcomes of their research and they are able to generate revenues, HEIs will have more incentives to create and commercialize knowledge. The Bayh–Dole Act enacted in the US in the 1980s is a classic example in this genre of policies. This act allowed the US universities to own the intellectual property generated through research funded by the state. While the jury is still out on the impact of the act on the generation of knowledge in HEIs (see discussion below), it is still seen as a potent instrument to incentivize HEIs to undertake research and commercialize it through different mechanisms (So et al. 2008).

Following the US example, a Bayh–Dole Act equivalent, the Protection and Utilization of Public Funded Intellectual Property (PUPFIP) Bill 2008 was introduced in the Indian Parliament on 15 December 2008 but is still to be passed. Universities and research institutions can patent the results of publicly funded research and academic inventors as well as academic institutions, and can get a share of royalties and licensing fees. The researcher (inventor) is to receive a minimum of 30 per cent of the royalties from the public-funded intellectual property (PFIP). However, she is restricted from publishing or disclosing the PFIP without prior notice of thirty days to the government. A violation will deprive the institution from future public funding and make it liable to repay the grant with interest. This is seen as a way to supplement the limited public budget for R&D.

There is, however, a lot of scepticism about the efficacy of such an act in the context of developing countries like India, which is very different from the US in terms of institutional context of HEIs and their research intensity. The quantum of inventive and innovative activity in India is rather low and few elite educational institutions are active in research (e.g., IITs, Institute of Science, Bangalore). Very few faculty or students opt for entrepreneurship and licensing options remain virtually unused. Besides, a large part of the research happens in the public sector, with limited commercial orientation; state-funded Council for Scientific and Industrial Research (CSIR) labs dominate in R&D and patenting activity. Technology Licensing Offices (TLOs) in CSIR labs and in other technology institutions are young and inexperienced. IP literacy is rather low although it is being built up in some institutions at a slow pace. Consequently, linkages between industry, government and academia are very limited. Finally, case studies have shown that IP has not been very critical for enterprise formation by educational and research organizations.*

The other reason why PUPFIP may not be very efficacious in India is that the educational institutional structure is very different in India compared to developed economies and even

---

* For example, see Basant and Chandra (2007b) for case studies on National Chemical Laboratory, Pune (a CSIR lab) and the Institute of Science, Bangalore and several IITs and IIMs.

many other emerging economies in East Asia. Apart from research capabilities, the ability of institutions to respond positively to an industry's production and innovation depends on the regulatory structure of the institutions of higher education. As mentioned in the last chapter, it has also been argued that the impact of any regulation of this kind or of changes in state funding for R&D would be positively linked to the autonomy of the HEI, its governance, and the competition it faces for research funding (Aghion et al. 2009). Unlike the US, the university system in India entails virtually no competition among universities and they have very limited autonomy. The structure of public research funding also does not create competition among universities. Competition among educational and research institutions, and within industry, stimulates collaboration between industry and universities. Institutional diversity and autonomy have played a critical role in UILs and technology transfers in US universities (Mowery and Sampat 2005; So et al. 2008; Mowery et al. 2001). Therefore, a sole focus on IPRs may ignore other types of linkages and roles of universities and more importantly, take the focus away from the restructuring of the university system that is essential to encourage collaboration with industry through various conduits.

Moreover, PUPFIP presupposes a *linear model* of technology transfer—patent, license and commercialization—and undermines the complex set of 'multiple reciprocal relationships' between university and industry which help build the innovation and entrepreneurial ecosystem. As discussed in Chapter 2, recognition of the complexity of relationships between different elements of the NIS is critical for the efficacy of the policy instruments targeting innovation. In all likelihood, given the limited empirical evidence discussed above, channels other than patenting and licensing may be more important for building these linkages.

Apart from the ground realities listed above, there are several other issues with the effectiveness of this IP-focused strategy. As Mowery et al. (2001) show, it is not entirely clear that Bayh–Dole had a significant effect on the content of research in American universities. Data on patenting, licensing and other types of commercialization show that the act did not facilitate technology transfer and commercialization in any significant manner (So et al. 2008). Furthermore, the ability of public-funded research to address questions of limited commercial value but significant public good may be compromised by reducing the difference between private and public research. This is especially the case in developing countries where public benefits and commercial advantages may diverge significantly because of the low incomes of large sections of the population. Interestingly, despite all these criticisms, there is talk of reviving the PUPFIP Bill in line with the recommendations of the new national IPR policy.* While the revival of IP-focused policy instruments is being discussed, there is no evidence of the impact of the existing policy of incentivizing commercialization of patents. In 2016, a patent box type of incentive was introduced wherein income in the form of royalties and technology licence fees received by Indian companies is to be taxed at a reduced rate of 10 per cent.†

---

* For details, see https://dipp.gov.in/sites/default/files/National_IPR_Policy_English.pdf

† The patent box policies, adopted in several nations, have been criticized as they are effectively seen as lower tax on 'monopoly' profits emanating out of a patent, which does not make sense. Besides, such tax benefits seem to have little positive effect on R&D investments (Mazzucato 2018).

Given these broad arguments, a more useful policy concern could be to ascertain how to create autonomy for HEIs and competition among them, while also facilitating building of networks and linkages with industry and other research organizations.

S&T developments and changes in global production and R&D networks are creating new opportunities for interaction between academic institutions and firms in India. Available data suggests that industry-academia collaboration in patenting and publication activity in India is very low; less than 1 per cent of patents filed and papers published were a result of such collaboration. The Government of India has set many schemes to incentivize academia-industry collaboration but details of ongoing collaborations under these schemes and their success are unknown.* However, there is some evidence to suggest that a few academic institutions in some Indian cities have utilized these opportunities to build linkages and interact with the city clusters in a rich variety of ways, primarily because only a few have the relevant knowledge to do so† (Basant and Chandra 2007a). There exists a hierarchy of institutions in terms of the strengths of their capabilities and linkages, viz., those that undertake only teaching (i.e., linkage through the labour market), those that do research and teaching and provide services such as testing (i.e., having linkages through all types mentioned earlier)

---

* Only about 0.4 per cent patents in India and 0.6 per cent publications were a result of industry-academia collaboration. More details on this and the list of the government programmes is available at: http://www.ctier.org/blog/2016/12/13/industry-academia-collaboration-india/

† In the case of the pharmaceutical-biotech industry, there are some indications that these linkages are beginning to form (Basant and Srinivasan 2016; Basant and Mukhopadhyay 2010).

and those that focus on specialized research (i.e., linkages that are predominantly driven by knowledge generation and dissemination). The distribution of HEIs across this hierarchy reflected low institutional capabilities and found that academic institutions rarely come together to advance these linkages collectively. However, a few who had some autonomy and capabilities seem to be gearing themselves up to participate in such linkages in a more systematic manner. The experience of institutions in Pune and Bangalore broadly supports the idea that institutions with research capabilities, given some autonomy, can co-evolve with industry to take advantage of emerging opportunities to build various linkages (Basant and Chandra 2007a).

Given all this, it has been argued that higher education has possibly become the weakest link in the innovation ecosystem in India. Apart from some of the issues identified above, what could have led to this decline in the quality of HEIs in India? In what follows, we discuss some of the issues that could have resulted in such a condition of the Indian institutions of higher education.*

4.3.1. Uniform Compensation and Automatic Promotion of Faculty

One of the reasons for all this could be the faculty reward system. Among the many shortcomings of the Indian faculty reward system is the low weight being accorded to industry interfaces, except perhaps in management education where faculty in some institutions get rewarded in various ways for industrial consulting and executive education. While it is still

---

* This is partly based on the issues earlier highlighted in Basant (2011).

debatable whether the creation of new *commercial knowledge* and ventures should be an important objective of the academic institutions, the Indian government is taking some intermediate measures to change this situation. Some elite HEIs are also enhancing their publications' focus to improve their rankings. The jury is still out on how research in HEIs can be encouraged to fulfil both the commercialization and publications functions, but some distortions seem obvious.

The University Grants Commission (UGC) specifies a salary structure for universities that receive its funding.* This makes compensation across different universities uniform in nominal terms, except for minor differences in allowances. Most other publicly owned academic institutions also benchmark themselves to this salary structure, which implies that faculty at these institutions face a uniform compensation structure.† In addition, most institutions typically follow a process of time-bound automatic promotion of faculty. This is a departure from the past when the number of senior positions was limited, and recruitment to these positions was by open competition, making it possible for faculty to be promoted early by applying for positions in other institutions, often in a different province. Consequently, mobility of good-quality faculty for early promotion through open competition is virtually non-existent.

A version of the automatic promotion policy was implemented way back in 1983. It has been subsequently

---

* For detailed information on UGC salary structure and recent changes, see https://www.ugc.ac.in/page/Pay-Related-Orders.aspx
† The Indian Institutes of Technology (IIT) and Indian Institutes of Management (IIM), central (funded by the federal government) and a few other designated institutions have a slightly higher salary structure than the normal UGC scale.

modified a few times but the essential features of the policy have remained the same. Seniority and time-bound promotion remains the essence of the promotion policy and research output was barely considered.* The UGC regulations notified in 2010 make some research mandatory with a very elaborate point system to measure the academic output of faculty for promotion purposes. However, both the quality and quantity standards are rather low and are unlikely to make any impact on the research output of the universities.†

Automatic promotion also blunts incentives to move between universities across the country. Everybody is quite comfortable in their own universities/states as automatic promotion is ensured. Consequently, a critical mass of good faculty cannot agglomerate, leading to a suboptimal dispersion of good faculty. Absence of research-oriented, internationally visible faculty groups results in low level of interest among the researchers looking for postdoctoral opportunities, which in turn adversely affects building a research ecosystem critical for productive research all over the world.

These promotion policies were essentially put in place due to the political pressures wielded by teachers' associations across campuses which were actively pursued by all political parties. The fact that education is a state subject in India has aggravated these political pressures, making resolution more complicated. The diversity among the faculty on campuses

---

* For details, see UGC Regulations (2000) for Appointment and Career Advancement of Teachers in Universities and Institutions affiliated to it.
† Reddy, Xie and Tang (2016) provide details of quality and assessment practices that leave much to be desired.

has declined dramatically in the last twenty-five years and one observes in-breeding as the dominant method of recruitment and promotion.

4.3.2. Separation of Teaching and Research

In the existing situation, teaching and research are increasingly not co-located because of the establishment of specialized research institutions. The establishment of institutions in the area of science and technology through the CSIR and the creation of social science research institutes through the Indian Council of Social Science Research (ICSSR) resulted in the exodus of 'research-oriented' faculty from universities to such institutions as the research facilities in these institutions were somewhat better. The consequent deterioration of the university system resulted in a self-reinforcing downward spiral that put the relationship between teaching and research further out of balance (Basant 2011; Krishna and Patra 2015). One of the key adverse impacts of such a separation has been that undergraduate teaching primarily undertaken by affiliated colleges has become almost completely de-linked from research (Yarnell 2006).*

Fundamentally, Indian universities are not involved in research, which is conducted by specialized science, technology and even social science institutions. In fact, only three Indian HEIs are ranked in the top 200 in the QS World University rankings of 2020, while China has more than five

---

* The data on publications in peer-reviewed journals by Indian universities and research labs also supports this contention. Overall, the research output is low and the distribution of publications is skewed in favour of a few research institutions (Basant and Mukhopadhyay 2010; Reddy, Xie and Tang 2016).

in the top 100.[*] This is not surprising as very few HEIs have good-quality physical and research infrastructure (Reddy, Xie and Tang 2016).[†] As noted, universities in India have very limited autonomy with respect to faculty compensation, which is largely uniform across universities. However, private universities and some specialized institutions can offer differential compensation, either in direct salary or in research support. Research-inclined students self-select into public or private labs (industry) or go abroad essentially due to differences in compensation (Banerjee and Muley 2008). In addition to the level of compensation, the lack of facilities and automatic promotion also diminish the research orientation in faculty. If faculty compensation and work environment is unattractive, quality erodes over time. This affects the very intellectual resource and capacity needed to generate new industry-relevant knowledge.

One of the most debilitating impacts of the automatic promotion and separation of research and teaching has been the non-availability of research-oriented faculty. As India tries to expand the base of higher education, both in terms of quantity and quality, the dearth of good-quality teachers is the most critical constraint. The politicization of the system has resulted in such a situation that any change is very difficult. Institutes in the fields of technology, medicine and management set up by the state, and which have done well, are being multiplied in such a situation. They struggle to

---

[*] The situation has been the same during the last five years. Interestingly, the rank of the University of Delhi, which is probably the best multidisciplinary university in the country, is higher than in the QS rankings.

[†] According to some evidence, ratio of PhDs to graduate engineers is less than 1 per cent (Banerjee and Muley 2008). This is sought to be changed in recent years.

find good-quality faculty and the remote locations of many of these institutions add to the problem in this regard.

4.3.3. Large University Systems vs Specialized Institutions of Professional Education

Recognizing the fact that not much research is happening in Indian universities, the Government of India created, through an Act of Parliament, two categories of HEIs—Institutes of National Importance (INI) and Institutes of Eminence (IoE). Both these sets of institutions have more autonomy in deciding their fees and the kinds of courses they can offer. Publicly funded IoEs will also get a large grant of INR 1000 crore. These HEIs are expected to have world-class standards, help make India the global knowledge hub and set benchmarks for excellence for other central and state universities. The main criterion for such a university was to be the quality and excellence of its research, recognized the world over. Both universities and specialized institutions of professional education in management and technology, etc., are part of these two categories of HEIs. The expectation also is that these institutions would help improve the record of Indian universities in world rankings.* In the same vein, the IIM Act of 2017 and IIT (Amendment) Act of 2016 were passed in order to provide these sets of institutions more autonomy and degree-granting privileges. The regulatory structure and rules of all these policy initiatives are yet to evolve and it is

---

* To the same end, the earlier government had announced the establishment of *innovation universities* in the country as a part of the XI five-year plan (2007–12). Foreign universities were also expected to participate in the creation of such universities. Not a single innovation university got established!

not entirely clear if the HEIs in various categories mentioned above will, in fact, have the autonomy that is claimed to be provided. Interestingly, none of the IITs, IIMs or any of the public institutions given the IoE or INI status are able to decide their salary structure.

The Human Resource Development (HRD) ministry of the Indian government has recently granted *IoE* status to three public and three private institutes, *enabling them to get full autonomy and special incentives to project them as 'world-class universities'*. This is part of the larger initiative undertaken by the HRD ministry to select twenty IoEs—ten each in the public and private sectors—which will enjoy complete academic and administrative autonomy. The IoEs will enjoy greater autonomy compared to other HEIs in terms of deciding the fees for domestic and foreign students, course duration, structure of foreign academic collaborations, and so on. The selected *public* institutions will also get INR 1000 crore (~$15 billion) each from the HRD ministry to achieve world-class status but no financial assistance will be given to the private institutions. This effort is similar to the ones tried (reasonably successfully) by China to upgrade their university system and also to the idea of creating 'innovation universities' by the earlier government which was not implemented. While the idea seems useful, it is not clear how it will actually get implemented. For example, despite the proposed autonomy, the salary structure of faculty in these IoEs will have to follow salary structures of the Government of India while there will be freedom to provide self-financed incentives.* In recent policy shifts, IITs and IIMs have been presumably granted

---

* The IoEs will follow the pay scales of that category of public institutions to which they belong. However, they may build in an incentive structure to attract talented faculty, with the condition

autonomy but they also do not have the power to decide their salary structures and have to deal with other forms of interference which creep in with political and bureaucratic hubris. Given this phenomenon, the ability to attract foreign faculty especially from the diaspora will remain problematic.[*]

While it is not clear how all these policy initiatives will pan out and if they would have the desired effect of making HEIs a vibrant element of India's innovation ecosystem, the fact remains that India has not been able to develop a single large multidisciplinary university that provides the scope and ambience for learning across disciplines, available in several US universities. Apart from separation of research and teaching, most professional education is provided through specialized institutes at both the undergraduate and postgraduate level, and these include elite institutions like IITs and IIMs.[†] The students in these institutions have very limited exposure to multidisciplinary learning, an essential ingredient for innovative thinking. The absence of a large university system and focus on specialized institutions of professional education has created a large gap in India's innovation ecosystem.

---

that the incentive structure would have to be paid from their own revenue sources and not from government funds.

[*] Economic Survey 2017–18 recommends that HEIs should attract back scientists from the Indian diaspora and refers to a variety of schemes that are available to use such opportunities but in the absence of decent salaries and a decent research culture, none of these schemes are likely to be successful.

[†] Many of these professional institutions may be affiliated to a university but can be independently or often concomitantly approved by the All India Council of Technical Education (AICTE) or other professional councils like the Medical Council of India. These have flexibility in designing their curriculum and student evaluation is done in-house.

In theory, many of the INIs and IoEs can potentially fill in this void by diversifying over time but that might take a long time. There is some scope for developing a network of educational institutions representing different disciplines that are co-located in a city cluster. Policy initiatives that can facilitate such city-specific collaborations can potentially create opportunities for multidisciplinary learning. This collaboration can take the form of research projects and joint offering of courses. There are possibilities of offering courses that combine technology, law, design, management and social science inputs. Some efforts in this direction might facilitate the co-evolution of networking institutions as envisaged in some versions of the triple helix literature.

We have suggested above that the strongest link between higher education and innovation-related activity in developing countries like India is likely to be through the labour market and it is this link that is of immediate concern to industry at this time. However, there are significant gaps here as well. The heterogeneity of industry's needs of (innovative) labour would require a response with heterogeneity at the institutional level. Based on international experience, Hatakenaka (2011) makes a very strong case for 'diversification' in the nature of higher education in India. But she also argues that diversity in quality standards across institutions needs to be explicitly recognized, instead of one standard fits-all perspective that the UGC currently has in terms of the uniform recruitment and promotion guidelines.* Besides, research-intensive multi-

---

* According to her, not all HEIs should be measured against a single set of matrices as excellence in vocational institutions demands different measures from those in academic institutions. For example, publishing in international journals may not be essential for all educational institutions (Hatakenaka 2011).

disciplinary universities should coexist with 'massification' of good-quality education which may require a rapid increase in good teaching-focused institutions offering liberal as well as vocational and professional undergraduate programmes. Finally, while one undertakes such a diversification strategy, there is a need to avoid the risk that the production of skills becomes too tailored to the current industrial structure which would limit the flexibility that is necessary in any workforce in today's globalized environment.*

In most developing nations in recent years, including India, governments are supporting the creation of incubators in HEIs to facilitate commercialization of university inventions through new enterprise creation. This is part of a significant policy push to create a vibrant entrepreneurial ecosystem to support innovation-driven start-ups. It is to the discussion of these endeavours that we now turn.

4.4. Creating an Ecosystem for Innovation-Based Entrepreneurship in India

India has had a long entrepreneurial tradition with more than 40 per cent of the non-agricultural workforce being self-employed. Policy-induced distortions to curb these entrepreneurial activities are being removed since 1991. Typically, traditional entrepreneurship works on family networks, internal funding and mentoring. This ecosystem is quite strong. However, the ecosystem for the first generation, independent, innovation-based entrepreneurship is still evolving and facing a variety of challenges. It is not entirely

---

* Wolf (2002) suggests that investment in basic skills such as mathematics or writing is maybe much more relevant than investment in domain-specific skills.

clear if the two systems can be made to interact synergistically. At the conceptual level, however, both can potentially benefit from the opportunities created by growth in the Indian economy. There has been significant start-up activity in India in recent years and a significant part of this vibrancy is due to first generation start-ups. Given this surge in the creation of new firms, India is being seen as a hotbed of start-up activity. According to some estimates, there were more than 20,000 active start-ups across various domains in India. New policies are being implemented to strengthen the ecosystem for the creation, sustenance and growth of start-ups. Supporting incubation is one such policy instrument and that has been a key policy instrument used by the state in India to create a vibrant entrepreneurial ecosystem in the country. Over the years, the focus of state-supported incubation activities has shifted from supporting rural livelihood-oriented small businesses run by less educated youth to the creation of urban-based, innovation-driven, high-value businesses run by well-educated youth (Sharma and Vohra 2020). Have these policies helped in building this ecosystem? In order to explore this question, we will need to start with some understanding of the features of such an ecosystem.

4.4.1. Elements of the Entrepreneurial Ecosystem

Given the discussion of NIS, UILs and the role of activities like incubation and acceleration in supporting technology-driven enterprises, one can think of various elements that need to be in place to support innovation-based entrepreneurship. Among others, these would include:

i. *Inventive activity*, by firms as well as universities (or with university participation through UILs).

ii. *Entrepreneurs* to take inventions to the market (innovative activity). These would include existing firms and new entrepreneurs including faculty, students and licensees.

iii. *Availability of finance* through various financial institutions including incubators, angel investors, venture capital firms, etc.

iv. *Diversity of skills* wherein capabilities relating to technology, design, management, entrepreneurship, etc., coexist or are 'connected' in some way. If one is considering HEI-based start-ups, this can be facilitated by availability of skills with different disciplinary backgrounds and fewer silos within the 'university' system that ensure interaction. In other contexts, it will require a 'market' for different capabilities that are provided by different entities through a network that is either geographically bound or dispersed.

v. *Mentoring support* especially for first generation entrepreneurs.

vi. *Institutions that bring all the above together* including incubators, accelerators, various types of industry and other networks (both formal and informal), clusters, state agencies and so on.

All these put together create 'entrepreneurship in the air' and associated flows of relevant knowledge. How has India done on these fronts? In what follows, we provide a broad perspective on some of these dimensions with respect to new enterprise creation in the context of the policies supporting incubation and acceleration activities in India.

4.4.2. Emerging Role of Incubators[*]

As discussed in the last chapter, HEI-based incubators are used as policy tools with the presumption that they take care of extant market failures and provide critical inputs for the formation of innovation-driven firms. The number of incubators in India has grown quite rapidly in recent years and it only lags behind the US and China in this respect. Conventional incubators are proliferating in India today. Virtually all well-known technology institutions have one and some of the management institutions are also experimenting with incubation. Roughly 60 per cent of the incubators are located in educational institutions, while the remaining are business-led. About 90 per cent of incubators in India are supported through various schemes of the government.[†] The largest supporter of incubation is the Department of Science and Technology (DST) through their programme of funding Technology Business Incubators (TBI). Most incubators run accelerators and about 60 per cent provide access to capital (Sharma and Vohra 2020). Unlike in western countries like the UK, access to funding is usually provided through seed funds operated by incubators in India and most of these seed funds are financed by the state. In the UK and elsewhere, incubators also facilitate seed funding but that is typically done through their network of angel and VC investors or

---

[*] The ideas discussed here draw heavily from a study that compared incubators in India and the UK (Basant and Cooper 2016).

[†] There are as many as twelve government bodies across several ministries with which these incubators are affiliated. Affiliation with different ministries at times results in a sectoral focus. Some estimates suggest that about 30 per cent incubators are sector-agnostic, while ICT (38 per cent) and agriculture (19 per cent) are sectors of focus (Sharma and Vohra 2020).

through funds provided by the university (Basant and Cooper 2016). This brings out the fact that incubation systems and early-stage financing are still to mature in India.

The models of incubation used by HEIs across nations vary a great deal and the efficacy of these models depends on a large number of factors (Basant and Cooper 2016). Although a large number of incubators in India are located in HEIs, unlike in the West, these are not part of large university systems or a larger geographically bound entrepreneurial ecosystem. They are typically located in HEIs focusing on technology, management or design education. As a result, while start-ups in technology or design institutions are often not able to get good business inputs, the ones in management schools face constraints with respect to technology and design-related inputs. In this sense, the 'structural' flaws of stand-alone specialized institutions get transferred to the incubators located in these HEIs as they are not part of a vibrant entrepreneurial ecosystem which essentially requires the existence of multiple disciplines. Besides, few Indian HEIs are active in research, fewer faculty members or students there opt for entrepreneurship and licensing options for university research remain virtually unused. As noted earlier, a large part of research activity happens in the public sector, with limited commercial orientation; state-funded CSIR labs dominate R&D and patenting activity. TLOs in CSIR labs are relatively young and inexperienced, and IP literacy is rather low. Consequently, linkages between industry, government and academia through the enterprise creation route are also very limited.

Broadly, the government-academia partnership has been a major contributor to the growth of incubation activities in India. This partnership has been more efficient than the state in creating its own incubation infrastructure (Sharma and Vohra 2020). In this sense, the government has partly been able to address the market failures in the entrepreneurial ecosystem and avoided government failures by not setting up and running the incubators on their own. However, the absence of large multidisciplinary universities where such incubators could be located and limited research undertaken by the HEIs has limited the impact of these incubators. Moreover, the state has not thought through the issues relating to the efficacy and sustainability of these incubators.

What do we mean by efficacy and sustainability of incubation in the context of HEIs? At the narrow incubation activity level, efficacy can be seen as successful commercialization of IP of the institution or others through enterprise creation by effectively supporting the start-ups with mentoring, capital, linkages, etc. Incubators located in developing economy HEIs can also create learning opportunities for faculty and students of the institution through live case studies, training, research and a connect with the entrepreneurial ecosystem. At a broader level, HEI-based incubation and acceleration activity can also contribute to building the entrepreneurial ecosystem in the region, as has been documented for institutions like Georgia Tech, MIT and Stanford.* A large multidisciplinary university is likely to be more efficacious in all these dimensions as compared to a stand-alone institution focusing on management, technology or design-related disciplines. Besides, an HEI that is financially and operationally autonomous, has an organizational design that provides incentives for its faculty and students and which is located in a geographical location having diverse capabilities

---

* See, for example, Youtie and Shapiro (2008) and Mian, Lamine and Fayolle (2016). Also refer to Chapter 3 for a discussion on this issue.

and a vibrant entrepreneurial ecosystem can benefit more from such opportunity.

But demands of sustainability may constrain these institutional choices.

What do we mean by sustainability of incubation in HEIs? Based on the experience of incubation in India, at least two interrelated dimensions seem important. One is *financial sustainability* of the incubator, which depends on the revenue streams it is able to generate. This is unlikely to happen if it is a continuous strain on institutional financial resources. Typically, rentals, mentoring fees and equity—the usual sources of revenue—are not enough to provide financial sustainability. The incubators therefore need to look for other sources of revenue like project grants and so on but those often do not provide long-term financial sustenance. The other can be termed as *organizational sustainability*. The incubator needs to attract and retain skilled and capable people for operating the incubator and providing effective incubation support. This, of course, requires financial resources. The other part of organizational sustainability comes from the organizational acceptance by the host HEI that having an incubator is a legitimate and useful activity. The HEI needs to be convinced that an incubator can potentially provide revenue streams through commercialization of innovations and learning opportunities and therefore a conducive organizational design is necessary.

Efficacy and sustainability of incubators are dependent on a variety of factors. In practice, efficacy and sustainability are intricately linked. If the incubation process is efficacious, financial and organizational sustainability of the incubation activity becomes easier. Conversely, if initial commitments ensure organizational buy-in and financial support, efficacy of the incubation process is facilitated. If the policy support assumes that the incubators would become self-sustaining in the short run and there is no need to finance them appropriately, such a policy at best would have limited success. Moreover, if the policymakers disregard the fact that diversity in an HEI or a cluster along with the autonomy of the HEI is critical for incubation activity to succeed, such policy initiatives are unlikely to succeed.

Several models of incubation are being tried in various technology institutions as well as management schools in India (Basant and Chandra 2007b; Basant and Cooper 2016). The organizational design of the HEIs is not appropriate for hosting and benefiting from incubators. Most of the HEIs do not have the legal mandate to incubate firms and hold equity in them. These issues have been typically resolved through the creation of a separate legal entity to hold equity on institutions' behalf. Faculty incentives for disclosure and commercialization through licensing and enterprise creation do not exist in most HEIs but are being put in place in some of them. However, significant gaps remain in terms of appropriate funding opportunities for start-ups supported by the incubators (see discussion in Section 5 below) and for financial sustainability of incubators themselves. As mentioned, most of these incubators are supported by state funds but such funding is not large and typically does not consider funding the salaries of good personnel with appropriate skills; the focus is essentially on creating physical infrastructure (Sharma and Vohra 2020).

Overall, therefore, systems are evolving for enterprise creation in HEIs to support innovation-based enterprises. Several models are being tried. It is a learning phase and institutional variations are very high. Fortunately, support for experiments to create an entrepreneurial ecosystem exists and it is exciting to see a variety of experiments under way. Creation of large multidisciplinary universities that exist in the Western economies does not seem to be on the cards in the short run, despite the state support for INIs and IoEs. As discussed, the idea of stand-alone incubation systems has some structural flaws which can potentially be corrected, at least partly, by networking and co-incubation in city clusters where institutions specializing in specific disciplines—science and technology, design, management, law, social sciences—coexist. One will have to think of an appropriate organizational design for such cooperation to succeed. Supporting incubation in a geographically bound cluster of HEIs can create economies of scope and scale. Diversification of some incubators into venture funding is likely to help in achieving financial and organizational sustainability. Some policy initiatives on the lines of the SETsquared partnership in the UK might be useful to explore for regions that have an agglomeration of good institutions in different disciplines.

4.5. Financing Innovation in India

It is evident from the discussion in Chapters 2 and 3 that capital requirements to finance innovation are not restricted to developing or acquiring an innovation (e.g., through R&D, purchasing technology or learning from others), but are also for taking it to the market. Such commercialization expenses include accessing various complementary assets (e.g., manufacturing, marketing and distribution, servicing) that can make innovation a fairly expensive activity. If the cost of capital is high, there is a disincentive to undertake innovation activities. This is particularly true in situations where technology markets are yet to mature as inventors cannot sell or license their inventions easily given the variety of market failures discussed earlier. Given this need for availability of financial capital at reasonable costs, capital market-related policies are critical for domestic firms as they respond to competition and exploit emerging learning opportunities. Cost of capital advantages of the MNCs can be a major barrier for appropriate response by domestic firms as they respond to enhanced competition through imports and FDI. Given the current problems with the Indian financial system, this may well be a major concern for adoption and adaptation of new embodied technology by Indian firms.

The cost of capital in India has been high. According to some World Bank estimates, the real interest for lending in India was in the range of -2.0 to 8.6 per cent during the period 2000–19 with an average of 5.1 per cent for the period as a whole. For the same period, the average rates for China, Russia, South Africa and the US were much lower at 2.0, -0.3, 4.4 and 2.9 respectively. Of course, the cost of capital in nations like Brazil was much higher at about 37 per cent.* One policy mechanism to ease the problems related to access and costs of capital for innovation activities is to provide research grants and subsidies and/or provide tax breaks on expenditures related to innovative activities (especially R&D). The R&D tax incentives in India, in the form of weighted deduction of in-house R&D, increased from 125 per cent in 1999–2000 to 200 per cent from 2010–11 till 2016–17. It was reduced to 150 per cent in 2017–18 and went down to 100 per cent in 2020–21 (Table 4.5). At 200 per cent, India had one of the most generous tax regimes (Mani 2018).

---

* These rates have been computed from the estimates available from the World Bank. (https://data.worldbank.org/indicator/FR.INR. RINR?locations=IN)

Table 4.5
**Policy Changes in R&D Tax Incentives in India
During Recent Years**

Union Budget	Change in R&D Tax Incentive	Scope
2009–10	R&D tax incentive extended to all industries in 2009–10	Scope of the provision of weighted deduction of 150 per cent on expenditure incurred on in-house R&D was extended to all manufacturing businesses except for a small negative list
2010–11	R&D tax incentive increased from 150 per cent to 200 per cent until 2016–17	Weighted deduction on in-house R&D expenditure increased from 150 per cent to 200 per cent
2016–17	R&D tax incentive progressively reduced from 200 per cent	Benefit of weighted deductions for R&D limited to 150 per cent from 1 April 2017
Since 2020–21	R&D tax incentive progressively reduced from 150 per cent	Benefit of weighted deductions for R&D limited to 100 per cent from 1 April 2020

Source: CTIER 2018; p. 15.

Availability of detailed information on the sources of finance for innovations in India and how it has changed over the years is virtually non-existent. But some data based on a large survey undertaken in 2013 shows that the bulk of the innovations in India seem to be supported by enterprises' own resources. Banks also play a significant role but the role of the government seems quite limited (Table 4.6). To what extent did the incentives and subsidies mentioned above incentivize firms to undertake R&D? While it is difficult to delineate the effect of tax incentives of this kind, preliminary findings suggest that the generous tax incentives did not have any significant effect on R&D expenditures of Indian firms (Mani and Nabar 2016). However, there is also some evidence to suggest that the weighted deduction policy positively impacted firm-level R&D investments by some SMEs implying that smaller firms do respond to this policy which may not be very useful for larger firms who would undertake R&D irrespective of the extant R&D tax regime (Nabar 2018).

Table 4.6
Sources of Funding for Innovations in Indian Manufacturing and Service Firms, 2013

Source	Percentage of reporting firms
Own resources	93.9
Banks	58.7
Government	8.4
International Organizations/ NGOs	4.7
Others	10.1

Source: Computed from World Bank Enterprise Survey—India (Innovation Module).
Sample size=3492.

One of the problems associated with benefiting from the R&D-related tax incentives is that the enterprise needs to be recognized as an R&D facility by the DST, Government of India. The eligibility criteria for such a recognition are onerous and transaction cost-intensive, making it difficult for relatively small firms to manage. Consequently, only a small proportion of firms take the trouble of applying for it.

Since firms without R&D units recognized by the state are not eligible for tax benefits, the incentives for them to report R&D expenditures are not very high. It is possible, therefore, that some of the R&D efforts remain unreported as they are either 'informal' or because accounting for them does not provide any benefits.

Another mechanism to fund innovation is to support R&D projects undertaken by non-state entities. A variety of policies have been tried in this space in India, including many in the PPP mode.* Since the initiation of economic reforms, the nature of R&D support programmes in India has undergone a change, especially in the context of SMEs. Two of these programmes—the Small Business Innovation Research Initiative (SBIRI) for early-stage funding of SMEs, and Biotechnology Industry Partnership Programme (BIPP) for larger and higher-risk projects—have received much attention recently. Modelled on the SBIR programme in the US, these initiatives are designed to incentivize firms to develop new products and processes through a matching grants system. A detailed review of these programmes by Aggarwal and Chawla (2014) has brought out some interesting insights. Two positive outcomes have been that the programmes have not only crowded in private investments, but have also facilitated building of linkages between businesses and academia. But the programmes can be improved on a variety of dimensions. Projects seem to suffer from bureaucratic delays, lack of operational flexibility, limited commercial orientation of experts, missing support services, weak commercialization linkages

---

* Aggarwal and Chawla (2014) provide details of these programmes which are managed by various departments of the Government of India.

and absence of a vibrant innovation ecosystem. Funding is often found to be inadequate to do the projects well and is often in the form of loans while the beneficiaries prefer grants. The government is also not able to ensure procurement, adding to market uncertainty. Moreover, since the research areas for which the proposals are sought are decided by the state, there are possibilities of government failure in that selection as well.*

It may be useful to create autonomy for the private players in such programmes through their involvement in selection of research areas, participation of industry experts in the selection of projects, and by tweaking procurement policies and linking the projects to incubators for commercialization and other support. Incubator support will be able to partly fill in the gaps in the innovation ecosystem through mentoring, venture financing, etc., provide some support services and facilitate commercialization. We shall revert to these options in the final chapter.

4.5.1. Financing Innovation-Based Start-ups

Funding of start-ups has picked up in recent years but it still has a long way to go. Angel investments have also grown but these investors have invested in only about 1831 deals during 2004–20. Since about 70 per cent of these deals happened after 2015–16, one expects this source of funding to grow more rapidly in the future.† But these deals have focused mainly on IT and ITES or mobile-centric start-ups;

---

* For more details on these issues, see Aggarwal and Chawla (2014).
† These estimates are compiled by the author from Venture Intelligence, India Database.

the share of such deals being as high as 74 per cent during 2004–20.*

Overall, the growth of Venture Capital (VC) and Private Equity (PE) funding (VCPE funding) has also been decent. But the quantum of investments has not been very large, especially VC investments. According to estimates available from Venture Intelligence, an amount of about INR 12.4 lakh crore was invested by VCPE firms in India during the period 1998–2019. Of this only about 8 per cent was in VC investments. The share of VC deals, however, was much larger at about 56 per cent. Within VC investments, the bulk (about 74 per cent) of them were early-stage deals but the share of early-stage investments was lower at about 58 per cent[†] (Table 4.7). The VCPE investment data also shows that there is a preponderance of IT, ITES or mobile-related deals that got funded by VCs in recent years. These segments constitute about 43 per cent of the deals and 32 per cent investment. Manufacturing has not been at the forefront of receiving VCPE investment. Taking the period 1998–2019 as a whole, the share of manufacturing in the total number of VCPE deals was only about 12 per cent while its share in investment was even lower at 9 per cent. The situation was worse when only VC investments were considered (Table 4.7).[‡]

---

* According to the Venture Intelligence database, the number of deals funded per year by angel investors increased from seven in 2004–05 to 278 in 2009–20. On average, only about fifteen deals were funded per year by angels during 2004–09, which increased to sixty-eight during 2009–14 and then to 235 during the most recent period of 2014–20. See Sabarinathan (2019) for similar conclusions.

† Early-stage investments are first/second round institutional investments into companies that are less than five years old, are not part of a larger business group and the investment amount is less than $20 million.

‡ An analysis based on somewhat older data shows similar trends (Dugar and Pandit 2017).

Table 4.7

Distribution of VCPE and VC investments by Industry Groups, 1998–2019

Industry Group	VCPE Investment			VC Investment		
	Per cent Share in Investment	Per cent Share in Deals	Average Investment per deal in (INR cr.)	Per cent Share in Investment	Per cent Share in Deals	Average Investment per deal in INR cr.
IT and ITES	32.2	43.2	89.7	57.7	61.3	16.6
Banking, Financial Services and Insurance (BFSI)	14.9	10.1	178.2	6.9	5.3	23.0
Manufacturing (MFG)	9.0	11.7	93.0	7.0	6.7	18.3
Healthcare (HC)	7.0	8.7	97.0	3.5	7.4	20.2
Engineering and Construction (E&C)	14.6	8.0	219.1	4.6	3.7	22.0
Telecom and Media (T&M)	6.9	3.2	258.6	2.3	2.3	18.0
Shipping and Logistics (S&L)	5.0	3.2	186.4	1.7	1.0	29.3

Non-Financial Services (NFS)	6.3	7.9	96.1	7.7	9.0	15.0
Others (OT)	4.1	4.0	121.7	3.5	3.3	18.5
Total	100 (1244320)	100 (10339)	120.4	100 (101311)	100 (5762)	17.6

Source: Compiled from Venture Intelligence, India Database.

Note: 1. Figures in parentheses are total investment (INR crore) and total number of deals.

2. Manufacturing includes Manufacturing, FMCG, Food and Beverages, Textiles and Garments; Healthcare comprises Healthcare and Life Sciences; Engineering and Construction comprises Engineering and Construction and Energy; Telecom and Media comprises Telecom and Media and Entertainment; Non-Financial Services comprises Advertising and Marketing, Education, Hotels and Resorts, Travel and Transport, Sports and Fitness, Other Services; Others comprises Agri-business, Retail, Mining and Minerals, Diversified and Gems and Jewellery. The classification is based on Dugar (2019).

While the number of early-stage deals that got funded had increased significantly in the 2000s (especially after 2009–10), the value of VCPE investments is significantly higher for late-stage deals (Figures 4.10 and 4.11). As expected, the average value of deals for late-stage investments is likely to be higher than early-stage ones. The data suggests that VCPE firms typically look at large deals as the transaction costs for them are rather high for smaller ones. Many incubators have access to seed funds (mainly given by the government) to provide the initial round of funding[*] but getting the second and subsequent rounds of funding is rather difficult. Thus, early-stage financing remains a challenge as early-stage funding is difficult to access.

Figure 4.10
Number of VCPE Investments in Various Stages (1998–2019)

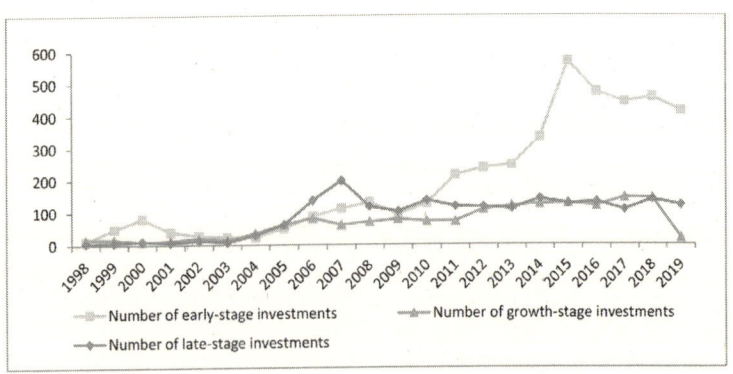

Source: Venture Intelligence Database downloaded on 22/06/2020.

Note: Figures of VCPE funding comprises early-stage, growth-stage and late-stage funding and VC funding of only early stage and growth stage.

---

[*] As mentioned, about 60 per cent incubators seem to provide access to capital in India (Sharma and Vohra 2020).

Figure 4.11
Value of VCPE Investments in Various Stages
(1998–2019)

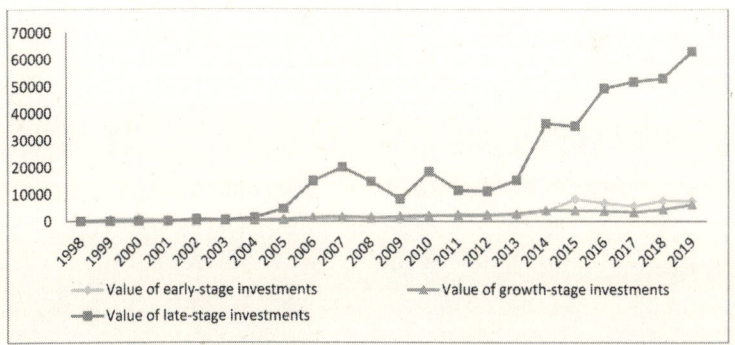

Source: Venture Intelligence Database downloaded on 22/06/2020.

Note: Figures related to value of funding are in INR crore. Here, VCPE funding comprises early-stage, growth-stage and late-stage funding and VC funding of only early stage and growth stage.

While capital market imperfections are quite important for incumbents, they are critical for start-ups. There remain significant market failures in early-stage funding of innovation-driven start-ups, despite the recent surge in policy interest and maturing of the VC industry wherein firms, till quite recently, behaved like private equity entities and provided little early-stage funding. A fund of INR 10,000 crore ($1.4 billion) was announced in January 2016 by the Indian federal government to be contributed for investing in start-ups. Through this initiative, the Government of India has created a Fund of Funds Scheme for Start-ups (FFS). The Small Industries Development Bank of India (SIDBI) manages the fund, and it supports different alternative investment funds (AIFs) registered with the Securities and Exchange Board of India (SEBI). AIFs extend funding support to start-ups (twice the SIDBI's contribution). This was to virtually transform the Indian VC industry and give a big boost to the Start-up India campaign as this amount was to be leveraged to raise more money from the private sector in the PPP mode. But the progress has been rather slow and the offtake of funds tardy due to a variety of issues. By 2019, only about INR 17 billion had been invested in 254 start-ups through SEBI's AIFs (David, Gopalan and Ramachandran 2020). As of July 2019, SIDBI committed about INR 31 billion to forty-nine AIFs registered with SEBI (Singh 2020).

While it is an interesting PPP model, a variety of operational and other issues seem to have hampered its efficacy. Tweaking of a few processes would help enhance the impact of this initiative. These may include bringing investing in start-ups and VC funds in line with tax benefits for investing in R&D, use of Corporate Social Responsibility (CSR) contributions for VC funds and so on.

4.6. Some Concluding Observations

In the context of the conceptual understanding of innovation, innovation processes and the role of public policy, this chapter explored the emerging innovation policy challenges for India. In recent years, contestability in the markets has been enhanced through a variety of policies, especially trade liberalization, abolition of industrial licensing and adopting a liberal policy vis-à-vis FDI. At the same time, Indian policymakers have experimented with a variety of other policies, especially the ones relating to the creation and support of innovation-driven start-ups. What has been the impact on innovation activities in India?

It has been shown elsewhere that the growth of total factor productivity (TFP) in Indian manufacturing during the post-reform period was the highest during 2001–09 when competition through import of final goods was growing and intensity of use of imported embodied technology (capital and intermediate goods) was also rising rapidly. The use of disembodied foreign technology through licensing also grew during this period albeit less rapidly (Basant 2021). As discussed, this was also the period wherein restrictions on entry, both for foreign and domestic entities, were being removed. So, an open economy seems to have helped the growth of TFP in the Indian manufacturing sector. Moving forward, tinkering too much with the trade policy or reversing the trend of liberalization by pulling back other liberalization measures are unlikely to help innovation-based competitiveness and may actually result in government failures as different sectors would lobby for protection of their markets from imports.

The post-reform period has seen an improvement in India in terms of its focus on innovation inputs and outputs. While both have increased in recent years, the change is neither dramatic nor consistent and we are still far behind countries like China. Increase in the private sector share in Indian R&D is noteworthy but it continues to be much lower than other nations. Overall, intensity of R&D and the use of foreign disembodied and embodied technology have shown an increase but here again, the changes are neither significant nor consistent. FDI inflows into India have increased quite rapidly in recent years and the nature of MNC involvement in India has moved up the value chain to some extent as many of them are undertaking R&D in India, often for global markets. However, the impact of FDI (MNC entry) seems to be mixed. While MNC R&D centres in India are an interesting development, the contagion effects of their activity are likely to be low as their research focus is more on global than local markets. Besides, the dominance of M&A-based FDI may not have resulted in significant inflow of knowledge into the Indian economy through MNC entry. Interestingly, foreign firms in India are not more R&D-intensive than domestic firms. MNCs do make local linkages resulting in knowledge flows, but anecdotal evidence suggests that China has been able to extract more out of FDI in terms of knowledge transfer than India due to a variety of conditions that are imposed on MNCs there.

A large part of the chapter focused on some of the key challenges in the domain of higher education. It was argued that while the labour market linkages between higher education and innovation are the most dominant in India today, there are significant gaps that need to be taken care of. In fact, there is an urgent need to strengthen these links as the higher order links of knowledge generation and enterprise creation would evolve gradually. Structural changes in higher education are required in order to make HEIs capable of responding to emerging innovation challenges and effectively play the role they are expected to play in the NIS. Higher education seems to be the weakest link in India's NIS. Given the issues of governance and other problems, the efficacy of the research done in HEIs is also rather low. Enhancing research intensity of HEIs is critical for building innovation capability and absorptive capacity, which, the chapter argues, is essential for any innovation-related policies to be successful. To face competition and exploit contagion conditions, firms will indeed build such capabilities, but their time horizons will be short and positive externalities would be a constraint on their efforts. Consequently, HEIs will remain critical to provide relevant capabilities on a large scale, keeping in mind the long-term needs of the economy.

While labour market linkages remain undernourished, the role of the university in innovation-based enterprise creation has slowly started to evolve in India. The review of enterprise creation models brought out certain gaps regarding the need for disciplinary diversity, sustainability of incubation activity in HEIs and early-stage funding. The issue of early-stage funding is particularly critical given the market failures in financing innovation in India and the high cost of capital in general. While a significant transformation of India's HEIs would be critical for innovation, it might be useful to explore policy innovations that can use the start-up funds to leverage public-private-academia partnerships apart from the PPP model that is currently being tried. The next chapter focuses on one such policy experiment of a public-private-academia partnership to address market failures in early-stage funding of innovation-driven start-ups in clean energy—Infuse Ventures. This experiment is particularly interesting as it not only provides insights on the mechanisms to address market failures in start-ups' activity in sectors with high social impact but also on ways to support the higher education system in becoming stronger as an integral part of the Indian NIS.

Annexure Table 4.1

Percentage Distribution of Patent Applications Filed under Various Fields of Inventions (1998–2017)

Year	Chemicals	Pharmaceuticals and Drugs	Food	Electrical	Computers/ Electronics	Mechanical	Biotechnology	General
1998–99	22.59	17.36	1.56	19.85	0.00	23.72	0.03	14.88
1999–00	18.40	21.91	2.34	19.22	0.00	26.01	0.20	11.92
2000–01	18.12	20.33	2.21	21.21	0.00	25.47	0.09	12.57
2001–02	18.34	20.72	2.59	17.23	0.00	27.67	0.05	13.41
2002–03	17.57	21.88	2.69	15.63	0.00	28.46	1.04	12.73
2003–04	23.40	20.02	0.98	16.85	0.00	21.54	0.18	17.03
2004–05	22.42	13.26	1.09	6.18	15.96	18.92	6.95	15.22
2005–06	23.71	9.02	0.41	5.20	23.26	19.32	6.22	12.85
2006–07	21.96	11.19	4.23	8.19	20.12	19.13	9.59	5.60
2007–08	19.08	12.77	0.70	6.61	14.49	19.23	5.84	21.28
2008–09	15.98	9.98	0.92	6.30	19.19	17.28	5.01	25.34
2009–10	17.54	8.95	0.80	6.93	22.30	19.76	3.80	19.91
2010–11	17.54	8.95	0.80	6.90	24.35	19.75	3.80	17.91
2011–12	15.51	6.39	0.68	9.63	9.78	22.49	1.82	33.69
2012–13	15.60	6.76	1.03	8.17	10.13	23.35	1.91	33.05

2013–14	15.76	5.84	0.90	10.18	10.27	26.35	1.51	29.20
2014–15	15.09	6.17	0.92	9.43	10.02	23.46	2.42	32.49
2015–16	13.78	6.32	0.83	8.75	12.77	21.67	1.89	34.00
2016–17	13.01	4.67	0.62	9.11	14.18	23.58	1.93	32.90
2017–18	13.25	5.73	0.72	8.94	12.72	24.18	2.07	32.38

Source: Computed from Annual Reports of Intellectual Property India, Office of the Controller General of Patents, Designs, Trademarks and Geographical Indication, Government of India, downloaded on 15/06/2020.

Note: Here, the General field comprises Biomedical, Biochemistry, Communication, Physics, Civil, Textiles, Metallurgy and Material Science, Agriculture, Engineering, Polymer Science and Technology, Microbiology, Agrochemical and Traditional Knowledge.

5

Financing Innovation in High-Impact Sectors

Insights from a Policy Experiment

5.1. Introduction

The last chapter highlighted various gaps in financing innovation-related activities in India. It also argued that incubation centres being supported in HEIs to foster innovation-driven entrepreneurship have several lacunae that need to be addressed. Apart from constraints on the availability of early-stage funding, these include the problems associated with the absence of a vibrant entrepreneurial ecosystem and sustainability of incubation activities in HEIs. Issues of funding and incubation of innovative start-ups become all the more complex in sectors that have very high potential for social impact but face a variety of market failures. A recent policy experiment—*Infuse Ventures*—undertaken at the Centre for Innovation, Incubation and Entrepreneurship (CIIE) at the Indian Institute of Management Ahmedabad (IIMA) addressed some of these issues in the context of the clean energy sector. The chapter provides an analysis of this experiment.

Clean energy is one of the most critical sectors for sustainability and social impact. Estimates suggest that an increase in the use of renewable energy will make a significant contribution to the growth of GDP (Inglesi-Lotz 2016) and have a positive impact on direct (Blanco and Rodrigues 2009) and indirect employment (Del Río and Burguillo 2009). There is also some evidence to show that clean energy-based electrification improves the living conditions of the community through improvements in the provision of health, education and other government services (Shoaib and Ariaratnam 2016). Moreover, renewable energy has the potential to reduce regional disparities by providing electricity in rural areas using local resources (sun, wind, biomass, etc.) which in turn create business opportunities (Nepal 2012).

Globally, markets for green products, processes and services are estimated to be very large and are expected to grow rapidly; the global volume of green-tech markets was about 3214 billion Euros in 2016 and is expected to grow to 5902 billion Euros by 2025 (BMUB 2018). The growing market combined with developments in technology creates opportunities for start-ups with innovative products, processes or services in clean-tech to establish and grow (Malen and Marcus 2017). Studies have shown that start-ups in the clean-tech space have a higher potential for innovation compared with all other start-ups and are more likely to use existing technologies in new ways apart from combining them with new technologies. Moreover, on average, clean-tech start-ups tend to introduce more new market offerings during their life cycle as compared to their peers (Loof et al. 2018).

Given all of the above and the environmental degradation and sustainability issues relating to conventional sources of energy, the need to enhance the footprint of clean energy cannot be overemphasized. However, the diffusion of such sources of energy has faced a lot of constraints and progressed at a slow pace across the world. One of the major constraints has been availability of capital for ventures seeking to commercialize new innovations in the market.

Infuse Ventures is a fund set up as a public-private-academia partnership (PPAP) to support innovation-driven start-ups in the clean energy space through early-stage funding. It is a policy experiment designed to address market and government failures that exist in developing economies like India in the provision of venture capital to start-ups in the clean energy space. An analytical description of this experiment is the focus of this chapter, and it includes details of the structure of the fund and its implementation. Through this description, an effort is made to identify useful insights on the policy initiatives and mechanisms to ameliorate government and market failures in supporting early-stage start-ups in the renewables sector. It is suggested that these insights can be leveraged to fine-tune policy initiatives to support innovation-driven entrepreneurial solutions in areas of high social impact like clean energy where technology dynamism is significant—say health and agriculture. It is also hoped that the discussion in this chapter will bring out sectoral specificities of clean energy that make the innovation support system in this sector different from others. Consequently, the analysis also provides some insights on the sectoral systems of innovation in a developing country context.

Infuse Ventures is analysed in the context of the market and regulatory gaps in the Indian clean energy sector which are addressed through its public-private-academia initiative. How the structure of the initiative takes care of the requirements of various stakeholders and strengthens the entrepreneurial ecosystem in the sector is also highlighted.

5.2. Market and Regulatory Failures and Early-Stage Investment in Clean Energy

Infuse Ventures was set up at a time (2013) when the VC industry in India was probably just past the infancy stage. It was growing but yet unproven, in terms of return potential, and undifferentiated, with no specialized sector-focused VCs. The last chapter has shown that despite the growth of early-stage funding in recent years, such funding remains difficult to get in India, especially for sectors outside the IT and ITES space. Early-stage investments in sectors like clean energy, which typically required large investments and have long gestation periods, were virtually non-existent. Even in more recent years, early-stage investments in renewable energy have been meagre.

5.2.1. Trends in Early-Stage Clean-Tech Investments

During the ten years before Infuse Ventures was launched in 2013–14, only thirty-one renewable sector deals were funded with about $190 million VC investment. This constituted only about 2.5 per cent of the total VC investment during that period with an average investment per deal being $7.3 million* (Table 5.1). Things have not improved much in recent years as the total VC deals in renewables have remained at around three per year. Moreover, the share of VC investments in renewables as a proportion of total VC investments also declined sharply from 2.5 to less than 1 per cent. However,

---

* Since investment amounts for all VC and PE deals were not available, the total investment is somewhat underestimated. The estimates of average investments per deal, however, consider only those investments for which such information is available.

the share of foreign VC investments in the deals has gone up and the focus has moved sharply towards solar energy-related investments.

Table 5.1
Venture Capital Investments in Renewable Energy (RE)—Patterns and Trends

Features	2004–05 to 2013–14	2014–15 to 2019–20
Total number of companies invested in	31 (3.1)	19 (3.2)
Total investment ($ million)	189.93	73.23
Percentage of RE investments in total VC investments	2.5	0.7
Average investment per deal ($ million)	7.31	4.31
Percentage of early-stage investments	74.19	63.16
Percentage of investments		
- By Indian investors	32.26	26.32
- By foreign investors	41.94	68.42
- As co-investments	25.81	5.26
- In wind energy	9.68	-
- In solar	45.16	94.74
- In hydel	12.90	-
- In biomass/biodiesel	9.68	5.26
- In others	22.58	-

Note: Estimates include investments in alternate energy deals. Figures in parentheses are average number of investments per year during the two periods.

Source: Venture Intelligence.

Private equity (PE) investments in the renewables sector have fared somewhat better. During the decade before the launch of Infuse Ventures, a total of 135 PE investments were made in this sector, the total investment being of the order of $3469 million and average investment per deal around $30 million (Table 5.2). While the average number of investments per year remained the same during the subsequent six years (~14), both the total investment and average investments have increased significantly in the more recent period. However, despite this increase the total investments remain meagre, constituting only about 5 per cent of the total PE investments. Moreover, the share of early-stage investments also went down from about 22 per cent to 16 per cent. Two other features of PE investments in renewables are similar to the trends in VC investments: (i) foreign investors have become more active in recent years, being investees in almost 78 per cent of the deals during 2014–20; and (ii) the share of solar energy in the renewable energy investments increased significantly and they contribute almost two-thirds of the total investments in recent years.

Table 5.2
Private Equity (PE) Investments in Renewable Energy (RE)—Patterns and Trends

Features	2004–05 to 2013–14	2014–15 to 2019–20
Total number of companies invested in	135 (13.5)	83 (13.8)
Total investment ($ million)	3469.41	7493.57
Percentage of RE investments in total PE investments	4.6	5.4
Average investment per deal ($ million)	29.65	107.05
Percentage of early-stage investments	22.22	15.66
- By Indian investors	28.89	15.66
- By foreign investors	56.30	78.31
- As co-investments	14.81	6.02
- In wind energy	24.09	19.15
- In solar	27.01	65.96
- In hydel	17.52	1.06
- In biomass/biodiesel	13.14	2.13
- In others	18.25	11.70

Note: Estimates include investments in alternate energy deals. Figures in parentheses are average number of investments per year during the two periods.

Source: Venture Intelligence.

Despite the potential, why have early-stage investments in the clean energy sector been lacklustre in India? What kind of market and regulatory failures contribute to this 'underinvestment'? Are there specific gaps in the Indian entrepreneurial ecosystem that need to be addressed in order to enhance early-stage investments in the clean energy sector? More specifically, why does the conventional venture capital model not work and why may direct intervention by the government be problematic?

5.2.2. Constraints on Early-Stage Investments in Clean-Tech—Government and Market Failures

In line with the discussion in Chapters 2 and 3, studies have suggested that 'push' and 'pull' factors play crucial roles in influencing green innovation (Loof et al. 2018). Cost savings and customer demand are critical pull factors that drive eco-innovation, while important push factors include technological development and capabilities. Given lumpiness of investment, long gestation periods and technological and market uncertainties, the introduction and diffusion of clean technologies is likely to be substantially delayed without government support in terms of taxes and R&D subsidies.

Following some earlier studies (IEA 2003; Owen 2006), one can argue that given the uncertainties associated with imperfect and asymmetric information about the renewable technologies, risk perceptions of producers, buyers and investors are high. Moreover, firms offering clean-tech options are typically not able to offer a competitive price as scale economies and learning benefits, which are available to incumbents using conventional non-renewable technologies, have not yet been realized by them. These disadvantages get accentuated by the fact that all costs associated with incumbent technologies may not be included in their prices as they may be subsidized, or regulatory distortions may result in situations wherein negative externalities caused due to the production and consumption of non-renewable energy are not internalized by the producers and consumers. At times, the cost of clean-tech may also be high due to the systemic gaps in existing infrastructure and capabilities that make the production and consumption of related products/ services cumbersome. This in turn increases the market power of the incumbents for whom all the systemic elements are in place, making entry of clean-tech start-ups difficult. High upfront investments, sunk costs and tax rules that require high depreciation rates add to the problems faced by clean-tech entrepreneurs. For clean-tech start-ups, one of the most significant impacts of these conditions is constraints on the access to funds. Such constraints are likely to be more severe in emerging economies where capital market imperfections are more pervasive. Meagre VC and PE investments in renewable energy endeavours, reflected in the data discussed above is, therefore, not surprising.

A recent study has strongly argued that the traditional VC model is the wrong model for clean energy innovation (Gaddy et al. 2016). The risk-return and time profile of clean-tech start-ups was not found to be amenable for VC investment unless these clean-tech enterprises were developing software solutions. This study of VC-funded start-ups across three sectors in the US (software technologies, medical technologies and clean-tech) showed that other than clean-tech software solutions start-ups, firms in this sector were poorly suited for VC investment because they: (i) required significant capital; (ii) had long development and commercialization timelines; (iii) were uncompetitive in commodity markets (presumably due to the price distortions mentioned above); and (iv) were unable to attract corporate acquirers providing exit to VCs. Gaddy et al. (2016) argue that given these specificities, the sector requires a more diverse set of actors and innovation models, and policymakers can support increased involvement of these actors in clean-tech companies by lowering the risk of investments in the sector. They specifically argue for public and private initiatives coming together.

What can the state do to address such capital market imperfections resulting in underinvestment by private entities in socially useful clean-tech innovation? Policy initiatives can take various forms to create access to early-stage finance, including exploration of direct grants, special funding, third-party financing options and so on. This will facilitate innovative new firms to introduce socially useful products and services in the market. Better access to funds will also lower entry barriers and create competitive pressures on incumbent firms, inducing them to innovate as well. It has been argued that since governments typically have a poor record of identifying the right kind of technologies, projects and/or enterprises, state support for such innovative ventures should not take the form of *direct* funding or grants, debt or equity financing by the government (Martin and Scott 2000).

How can we effectively use public funds to provide investment funds in a risky environment fraught with different types of market failures and avoid government failures as well? Effectively, one needs a policy instrument that leverages the investing and monitoring ability of a VC firm without imposing complicated contracts and/or bureaucratic administrative mechanisms. An appropriate PPP can potentially do the trick. PPPs provide an attractive investment mechanism for governments as such arrangements entail the use of private funding as well as expertise to satisfy public needs. At the same time, the private sector gets the opportunity to mitigate some of their risks. In fact, PPPs are deemed to be one of the best mechanisms to overcome the government's budgetary constraints and share investment risks between the public and private sector. Cedrick and Long (2017) document a variety of physical projects across nations in the renewable energy sector that have been undertaken under the PPP framework. Apart from power producers and developers, a variety of private sector financial institutions (banks, pension funds, etc.) have participated in these projects but VCPE funds have usually been absent in such arrangements. Various types of financial arrangements are observed in these projects. The governments typically contribute through grants, concessional loans, guarantees, public equity, forex risk mitigation, mezzanine funds and/ or subordinated equity. The private entities provide funds through various types of debt (loans, bonds, credit lines, etc.), equity, mezzanine funds and credit guarantees.

While Infuse Ventures is a PPP (or PPAP) endeavour, the partnership is not to undertake an identified physical project but to create a fund that will invest in early-stage innovation-driven clean energy start-ups and thereby address some of the market and government failures enumerated above. In this sense, it is significantly different from physical projects undertaken through a PPP mode which may not face the same kind of risks as the start-ups. Consequently, the role of financial institutions engaged with early-stage investments is expected to be higher here. However, the core idea is to mitigate the constraints that result in underinvestment in clean energy projects by the private sector.

The PPP in Infuse Ventures is not only different from other PPPs with respect to the dimensions identified above, but on other counts as well. While some of these will be discussed later, two of them are highlighted here to provide a context for the discussion in the next section: (i) an academic institution (IIMA through CIIE) is part of the arrangement (which makes it a PPAP); and (ii) the diversity of stakeholders in the arrangement is very high. This raises a variety of questions. Is there an advantage in having an academic institution or an incubator as a stakeholder in such arrangements? Is diversity among stakeholders useful? What incentives would work for different stakeholders? Can these incentives work at cross-purposes?

5.3. Setting Up Infuse Ventures: Origins and Stakeholders

Making a commitment to alleviate climate change, the Government of India had introduced the National Action Plan on Climate Change (NAPCC) on 30 June 2008.* It comprised multiple national missions that collectively focused on promoting understanding, adaptation and mitigation of climate change as well as energy efficiency and conservation of resources. The Jawaharlal Nehru National Solar Mission (JNNSM) was launched in 2010 to drive growth and development of the solar energy sector in

---

* The National Action Plan on Climate Change (NAPCC) encompasses a range of measures. It focuses on eight (only two are mentioned here) missions, which are as follows: (a) National Solar Mission: The NAPCC aims to promote the development and use of solar energy for power generation and other uses, with the ultimate objective of making solar competitive with fossil-based energy options. It also includes the establishment of a solar research centre, increased international collaboration on technology development, strengthening of domestic manufacturing capacity, and increased government funding and international support. (b) National Mission for Enhanced Energy Efficiency: The NAPCC recommends mandating specific energy consumption decreases in large energy-consuming industries, with a system for companies to trade energy-saving certificates, financing for PPPs to reduce energy consumption through demand-side management programmes in the municipal, buildings and agricultural sectors, and energy incentives, including reduced taxes on energy-efficient appliances.

India as part of the NAPCC. The Government of India had also created an environment conducive to investment from foreign investors by not only allowing 100 per cent FDI in the renewable energy sector but also encouraging build-own-operate (BOO) projects.

While the government had taken multiple steps to promote grid technologies at the higher level, it had been looking for ways to promote smaller, modular renewable applications like that of the solar water heater, the cooking stove, etc. When the Government of India's Ministry of New and Renewable Energy (MNRE) was in the process of announcing JNNSM, it was felt that entrepreneurs could play a vital role in helping the government achieve the targets being set for JNNSM. They had an existing relationship with CIIE (see below), and felt that the initiative could build upon these past efforts done by CIIE and MNRE.

CIIE has focused its activities in areas where market failures are high and where the gaps in the start-up ecosystem are critical. It was in this context that the CIIE team started to explore an opportunity for catalysing the clean-tech space in India and approached MNRE to explore possibilities of collaboration. While the sector was largely underserved in India, the clean-tech space was gaining visibility in the eyes of the government; new schemes were being designed and implemented to support entrepreneurs in the sector. As one of India's first incubators to develop interest in the clean-tech sector, CIIE had developed a strong partnership with MNRE. This led to several 'acceleration' initiatives in collaboration with the MNRE, such as involvement with the Solar Innovation Programme and the Renewable Energy Search Programme. These were, however, one-time programmes and involved only CIIE and MNRE. It was felt that these isolated acceleration-focused initiatives were not enough and India needed stronger long-term initiatives to make an impact. Building a full-fledged clean-tech ecosystem, which could effectively promote innovation-driven start-ups needing large investments, was seen to be critical. CIIE's own experience was that several promising start-ups that it supported through seed funds struggled due to the non-availability of subsequent rounds of funding for scaling up. As discussed earlier, VCPE firms were not very active in providing early-stage financing to clean-tech firms. It was in this context that the idea of setting up a clean-tech venture fund was germinated.

As mentioned, clean-tech globally has been a very risky sector. Running this fund would require a lot of resources—financial as well as non-financial. However, in the face of a slowly growing clean-tech space, developing an ecosystem for such start-ups held enormous appeal and CIIE was definitely not the first institution to understand this opportunity. Moreover, several instances to establish something similar had failed. For example, the India Climate Innovation Centre model was an attempt by the World Bank to replicate clean-tech ecosystems created across the world—one that was planned in its entirety, but failed at the execution stage due to incompatibility with the Indian context. Similarly, International Finance Corporation received a commitment of over $60 million towards setting up a dedicated clean-tech catalytic investment facility for investing in early-stage ventures across the world, but eventually disbanded the line. Their original effort was to host it in a government body (like the Ministry of Power's Bureau of Energy Efficiency or DST's Technology Development Board), but finding a government champion was hard.

5.3.1. Centre of Excellence in Clean-Tech and Infuse Ventures

How was CIIE's experiment made appropriate to deal with the unique vagaries of the Indian context? CIIE's work in the clean-tech space was slowly gaining recognition. The opportunity to start creating the ecosystem that had been in the pipeline for CIIE gained significant impetus when CIIE was approached by BP* to do a landscape study of the clean technology space in India. As a private entity, BP's motives were clear: BP Ventures, the investment arm of BP, was on the lookout for new game-changing technologies, and partnering with CIIE would allow them to create a presence for themselves in clean-tech in India. It was also at this time that the discussions with MNRE for supporting a Centre of Excellence (CoE) at CIIE for renewables and clean-tech gained momentum, a centre that would help create a vibrant ecosystem for start-ups in this space. MNRE mandated CIIE to raise matching capital for the CoE and its start-ups through co-investment and follow-on investment by other external investors. Given its understanding of the gaps in the early-stage financial markets for start-ups, CIIE decided to use the VC fund path to crowd-in capital and build a portfolio of companies that would attract further private sector validation. The route of CoE was found to be most appropriate as it not only facilitated the participation of the private sector but also made MNRE's own participation possible. It could support the CoE with a *grant*, part of which could be leveraged by CIIE to seed a fund as MNRE could not directly invest in the fund. However, while raising capital and fundraising from other investors would take a lot of time, CIIE initiated the

---

* Earlier known as British Petroleum.

activities of the CoE by directly supporting start-ups through grants and seed investments.

The ongoing link with BP meant that BP became the natural first partner for CIIE and the idea of Infuse Ventures—to help accomplish the objectives of the CoE—was born. The core idea was to set up a fund in which CIIE could partner with both the government as well as the private sector. Each stakeholder would be an investor in the fund, with its operations being the responsibility of CIIE and the fund being used to support early-stage start-ups in the clean-tech space. Infuse Ventures, therefore, was essentially seen as an initiative that would actively create an ecosystem in the clean-tech space by searching for and supporting a variety of opportunities through several mechanisms including grants and equity investments. The fund was expected to take equity stakes and operate on a commercial basis with a profit-maximizing objective, but would have a sector focus on start-ups that could have a material impact on India's long-term clean energy markets and strategy.

While Infuse Ventures was envisaged as a part of the larger initiative to set up the CoE, CIIE was aware that it was critical for creating a larger ecosystem. Even Infuse benefited from this larger initiative through the creation of a pipeline for start-ups that could be funded. Right from CIIE's early work in the clean-tech sector it was clear that an Infuse-like intervention would have to be focused on seed and early-stage start-ups because a strong pipeline did not exist in India. Apart from VCs, angel investors, incubators and accelerators were also focused on consumer tech and clean-tech entrepreneurs were looking to get off the ground.

In line with the broad objective of the CoE to help create an ecosystem for the clean-tech start-ups, a multi-pronged synergistic approach was adopted having four core elements:

i. Scouting and accelerating new ideas through training and boot camps which was done through a programme called PowerStart;

ii. Provision of pilot grants for prototyping and seed capital through a seed fund (Renewableseed);

iii. Provision of scale-up and follow-on capital through the equity fund, Infuse Ventures; and

iv. Policy advocacy and support for government interventions that could catalyse entrepreneurial activity in the clean energy space.

PowerStart was essentially an accelerator programme for very early-stage clean-tech start-ups. A week-long capacity-building boot camp programme provided tools, skill sets, networks and know-how that could help start-up teams develop a road map for the next four to six months. Top teams received ongoing mentorship support to implement their plans and companies that implemented successfully had the opportunity to pitch their business to Infuse for funding. This boot camp and mentoring were also available to aspiring entrepreneurs (entrepreneurs in residence) who wished to explore new opportunities and build businesses from scratch over three to six months. Successful entrepreneurs could also apply for seed funding from Infuse as their engagement progressed.

Renewableseed (REseed) programme provided prototyping or piloting grants of up to $20,000 as grants for clean-tech entrepreneurs trying to develop a new product or piloting it with a customer.

Infuse could provide up to $2 million in early-stage investment to start-ups that had piloted their solutions and achieved a scale where Infuse support could help them further commercialize their offering. Earlier, CIIE could only provide seed funds of the order of $1,00,000 but its start-ups often struggled to raise capital even when they were doing well, for reasons discussed above and in the last chapter.

All these synergistic activities were designed to help build an ecosystem to nurture new ventures in the clean-tech space by providing knowledge, mentorship and capital at different stages of the venture and policy insights. At least two aspects of this initiative were unique and need a mention. One, while each of the four activities has been tried in different parts of the world individually, presumably for different sectors, this was probably the first initiative undertaken by an academic incubator like CIIE, which combined all the activities. Second, this was the first case of an academic incubator setting up a fund to provide scale-up capital through a PPAP. The Infuse Ventures model held great appeal. It was expected that the backing of the government to lessen the burden of regulations, provide credibility, reduce risk of the private sector and the technical and managerial knowledge offered by CIIE as well as the private sector, would help budding start-ups perform well.

5.3.2. Diversity in Contributors to Infuse Ventures and Other Stakeholders

To set up the fund, a sum of INR 23 crore was given to CIIE as a grant by MNRE. This amount was to go as CIIE's contribution to the fund, which it would leverage to raise at least an equal amount from the private sector before the fund could be utilized for investment. CIIE was able to raise much more and the total fund was of the order of about INR 107 crore in which nine investors contributed (Table 5.3). The contributors (stakeholders) differed a great deal from each other in different respects. Broadly, they can be divided into three categories:

a. *Developmental* – Contributors looking for a developmental impact. These included MNRE, TDB, ICICI, IFC and SIDBI
b. *Strategic* – Contributors who had a commitment to sustainability but were looking for strategic opportunities through the fund for investment and/or acquisition. Godrej and BP fall in this category; and
c. *Commercial (Financial)* – Contributors who were looking for good financial returns. The two banks, UBI and BoI belong to this category.

Table 5.3
Investors in Infuse Ventures and their Contribution in the Fund

Investor	Type of Stakeholder	Effective Commitment (INR crore)
Ministry of New and Renewable Energy, Government of India (CIIE)*	Developmental	23.00 (21.5)
International Finance Corporation	Developmental	16.01 (15.0)
BP	Strategic	15.00 (14.1)
Technology Development Board, Government of India	Developmental	10.00 (9.4)
Bank of India (BoI)	Commercial	10.68 (10.0)
ICICI	Developmental	5.00 (4.7)
SIDBI Venture	Developmental	16.01 (15.0)
Godrej	Strategic	6.00 (5.6)
Union Bank of India (UBI)	Commercial	5.00 (4.7)
Total		106.75 (100.0)

Note: Figures in parentheses are the percentage share of each contributor/investor.
*The MNRE contribution was made through CIIE initiatives.

Within the developmental category, MNRE and TDB were Government of India entities focused on the development of technology. SIDBI and ICICI were venture capitalists, the former being state-owned and the latter in the private sector. IFC being an international financial institution, part of the multilateral system and focused on private sector development in developing economies, brought in an international flavour, including standards, to the group. In the strategic group, while BP was a large international conglomerate with significant investment in clean energy-related R&D, Godrej—a large domestic corporate entity—also had significant interest in this space. The two domestic banks in the third category were not very different from each other.

Overall, therefore, there was significant heterogeneity among the investors and Infuse Ventures was a combination of 'developmental' and 'financial' capital. While all investors knew from the beginning that they could not expect financial returns anywhere close to that of VCs, the expectations did vary across contributors. For MNRE, its contribution was a grant and they were looking for the leverage they could provide to 'crowd-in' private capital into the sector. Developmental investors like TDB and SIDBI joined in with the intent of catalysing technology development and start-ups/SMEs, respectively. IFC was in the process of setting up a clean-tech-focused corpus of around $60 million and an investment in Infuse was a pipeline development effort.* Godrej and BP were also looking for some strategic returns and the two banks probably had the highest expectations in terms of returns.

The fact that Infuse Ventures was part of a larger endeavour to build an ecosystem for clean-tech start-ups and the contributions to the fund included both developmental

---

* This plan by IFC was subsequently discontinued because of the risks involved.

and financial capital meant that structurally the initiative had multiple objectives. It could not be operated as an entity that essentially focuses on financial returns as ecosystem building and development impact were also part of the mandate. Infuse Ventures, therefore, was not a typical VC entity and managing it required managing a variety of stakeholders with somewhat different objectives.

For CIIE, management of stakeholders was not restricted to managing contributors to the fund with divergent expectations. It also had to deal with multiple other partners who joined the endeavour to supplement efforts in the three activities other than the provision of scale-up capital through Infuse. For example, the CoE was not only expected to facilitate access to commercial capital which entrepreneurs need to scale up their solutions but also provide training and hand-holding support. Part of the money that was given to Infuse by MNRE was set aside for this purpose but was also topped with the support from Asian Development Bank (ADB) as a grant infusion.

Broadly, contributions from ADB/UNEP through Seed Capital Assistance Facility (SCAF) significantly complemented the funds from MNRE to support REseed in providing prototyping and piloting grants. This helped create a pipeline for investments by Infuse Ventures. In addition, the ADB grant supported the PowerStart programme and the provision of technical assistance (TA) support for pipeline and portfolio companies. This funding allowed the CoE to provide up to 10 per cent additional grant to start-ups in which Infuse had made investments. Once again, the ADB/UNEP support was of the developmental capital variety and since it was playing a significant role in creating a pipeline for Infuse as well as funding TA support to portfolio companies, issues relating to 'mixing' of developmental and commercial capital were not restricted only to Infuse activity, but effectively pervaded a large part of the ecosystem-building endeavour of the CoE.[*]

5.3.3. CIIE as a Key Stakeholder

So far a variety of 'external' stakeholders have been listed. But it needs to be explicitly recognized that CIIE at IIMA is also a significant stakeholder in Infuse Ventures and its own objectives and capabilities need to be understood. With a mandate to support innovation-driven start-ups with significant impact, it has been a very successful incubator. It potentially had access to the networks of IIMA. Over time, it had built a lot of experience of incubation and was the first to launch an accelerator programme in India and it had the ability to provide useful business inputs. It had also worked with several other incubators, both domestic and international, to help build their systems. These efforts included programmes focused on clean energy.[†] However, its investment experience was restricted to seed funding which did not exceed INR 50 lakh per venture. The monies it had raised were typically grants from the government and foundations. It did not have any experience of raising financial capital;

---

[*] In addition to what has been stated above, CoE also raised resources to undertake related activities. For example, Rockefeller Foundation provided support for the catalytic capital facility with cKinetics. This effort focused on mini-grid and distributed renewable energy companies to bridge the debt gap and took the form of soft debt.

[†] For example, CIIE joined hands with Irena (International Renewable Energy Agency) and ran a couple of programmes focused on training aspiring incubators across Africa and south-east Asia. It has provided training to a dozen-odd incubators.

even for seed funding, the main sources were developmental capital. Consequently, CIIE had neither raised nor deployed commercial capital *before* Infuse Ventures was set up. But as a part of this endeavour, it was not only expected to raise both kinds of capital but also invest in a synergistic manner so that the objectives of different stakeholders were satisfied.

Moreover, unlike incubation labs present in academic institutions with significant technology focus across the world (such as MIT or IITs in India), CIIE did not have the technical expertise to support highly tech-based start-ups through adequate technical guidance. But given its own credibility and reputation along with that of its parent institution (IIMA), it was feasible to get technical advisers from outside. Besides, CIIE could leverage its unique association with the government to insulate start-ups from the vagaries of changing regulation.

Finally, the context of CIIE also needs some elaboration in order to draw some generalizable lessons from the Infuse experience, especially in the context of the discussion on incubators in the last chapter. Usually, as mentioned earlier, incubators in academic institutions across the world do not make money except when some of the start-ups they have supported and seed-invested in become enormously successful. That rarely happens and when it does, it takes a fair bit of time. Therefore, the business model of an academic incubator is essentially grant-based as they are usually not 'profitable' entities. Of course, they can play a significant role in facilitating the technologies developed in the university to the market and in that sense play a critical role for the institution. How do incubators like CIIE achieve financial sustainability? Since CIIE is not part of a technology institution it cannot be seen as performing a larger role of commercializing technologies developed in-house. Participation in Infuse marks a major transition for

CIIE—from an incubator-accelerator to an entity that combines incubation-acceleration with VC investment activity.* If this transition is successful, the issue of financial sustainability of CIIE also gets partly resolved as the management fee of about 2.5 per cent that Infuse Ventures provides becomes a stable source of funding during the period of the fund. From that perspective, the transition was useful for CIIE but no other academic incubator had made that transition and on the capabilities side, the Centre could be seen as somewhat deficient! Effectively, Infuse created an opportunity for CIIE for a significant organizational change which in turn created a need for a different mindset. But, given its experience and capabilities, was CIIE an appropriate anchor for Infuse?

5.3.4. Managing Stakeholders

As mentioned, bringing together different types of partners with different expectations was a critical task for CIIE to perform as the anchor of the project in order to make it a success. Commercial investors expected a good Internal Rate of Return (IRR), the government and the development organizations looked for high impact in terms of creating new jobs, markets or enterprises, and corporates sought strategic opportunities. A balancing act had to be done as CIIE pursued multiple objectives simultaneously with different stakeholders pulling in different directions. For example, the government preferred impacting more start-ups (a.k.a. numbers) while a spraying of capital would not have the right outcomes

---

* Interestingly, a similar transition has been made recently at MIT through MIT Engine. We will get back to this transition in the concluding section (see https://news.mit.edu/2016/mit-announces-the-engine-for-entrepreneurs-1026).

financially. Similarly, while the government wanted Infuse to take more R&D risk, Infuse Venture's close-ended nature as a fund meant it could not back long-gestation projects.

Some partners required significantly more paperwork than others and as a consequence, the administrative time required to follow specific protocols that varied significantly across partners was very high. A full-time person essentially played the role of reporting to various stakeholders. The activities and outcomes relating to Infuse had to be 'translated into different languages' to satisfy the reporting needs and expectations of the partners.

One factor that seems to have helped in managing partners is that with several of them, CIIE had prior work experience. The Centre also had significant experience of dealing with developmental capital provided by a variety of donors—state-supported and private, domestic as well as international.

As mentioned, CIIE had worked with MNRE on several programmes before Infuse was launched. The seeds for Infuse were actually sown during this period. This engagement was quite deep as it also involved CIIE actively working with MNRE to grow the market through policy interventions. For example, when the Government of India was setting up the National Solar Mission, the scheme being created was not start-up-friendly and was meant to support only large operators as limitations were being put on the financial and technical history of the start-up to avail subsidies, etc. CIIE engaged with the Government of India for changes in policy to make it start-up-friendly, became a programme administrator for the National Solar Mission and during the early days of the Mission, many start-ups actually availed of the subsidies through CIIE. Similarly, in specific segments like solar water pumping, CIIE worked closely with the government and made sure that there were relevant provisions created for start-ups. CIIE joined hands again with MNRE and various other government departments to overcome other policy hurdles that existed for start-ups. This policy engagement with MNRE and other state organizations continued after Infuse was set up.* It needs to be recognized that CIIE's earlier association with MNRE and other parts of the government facilitated a significant mindset change at their end. Since Infuse was essentially a risk fund for the government, it required an enormous amount of convincing the internal finance team to approve the proposal. The finance team was more used to approving projects with tangible outcomes like capacity building or asset generation; building an ecosystem was not one of them. Moreover, the ministry had to formally buy into a situation wherein only 20–30 per cent or less of the supported start-ups would attain sustainability.†

The senior team at MNRE had realized the importance of a fund like Infuse to support ecosystem building in the clean-tech space. They were also convinced that more than the amount

---

* Similarly, a geo-cooling company, which had created a fairly breakthrough innovation that leveraged lower ambient temperature that exists under the ground to bring about cooling of large spaces, faced peculiar regulatory hurdles. For every project when they undertook drilling, they needed to go back to the Government of India *ab initio* and get several layers of approval as the laws were archaic and had not envisaged such possibilities when these were created. CIIE was able to facilitate some of those approvals through regulatory changes.

† There is also a potential issue of the dynamics within the ministry: the flexibility that is available at different levels of hierarchy and if the senior management is willing to use this flexibility to undertake policy experiments; what programmes/budgets are in place that can be leveraged to experiment. Interestingly, the money came from the solar mission budget although it was to be used in multiple clean-energy spaces.

being contributed to the fund, the presence of the government as a partner on this initiative would help attract other private players to this space. Eventually, MNRE—along with the Planning Commission—decided that the work proposed to be carried out by CIIE under the project was of a developmental nature, and therefore decided to provide a grant-in-aid to help develop CIIE as a CoE. Part of the grant was to be invested in the fund as CIIE's contribution and the CoE was mandated to raise matching funds from other sources to provide financial and mentoring support to start-ups in the sector.

While the government was a critical investor for Infuse, CIIE was aware that significant involvement of the government could pose other problems. Besides, there were issues relating to the continuity of MNRE support. CIIE had complete support from the current Secretary and Joint Secretary at MNRE, but it was unclear if future representatives were likely to view Infuse as favourably as the current dispensation did. It is well known that changing representatives often leads to a lack of continuity, affecting current and future funding and support. This did happen to some extent and a significant amount of time had to be spent to bring new representatives up to speed with the CoE's activities.

Another concern was that frequent policy changes could lead to a change in the goals and yardstick for measuring success at the government level. These policy changes also meant that the start-ups commonly faced the 'moving goalpost' problem in this space. Infuse could leverage its unique association with the government to insulate start-ups from such severely changing goals. Therefore, the level of involvement with the government was a critical facet of Infuse and important for CIIE to manage well.

Other non-financial investors like IFC or BP also had a changing internal mandate with respect to clean-tech, and the strategic value derived from their participation was lower than expected. However, their participation in the fund itself provided a significant signalling advantage. More active participation from them would have enhanced the strategic advantage they could provide to the fund as well as some of the portfolio companies.

Apart from the corporates that had invested in Infuse, CIIE also engaged with almost twenty–twenty-five other Indian and MNCs, to help its start-ups grow both through business linkages as well as financial investments. In the absence of adequate interest from VCPE investors, such close linkages helped de-risk the activities of the start-ups.* Many portfolio companies derived enormous strategic value working with corporates. With CIIE's intervention, corporates understood the innovation better and consequently the start-ups were able to leverage the resources of the corporates—capital, distribution channels, production expertise, etc.—to scale up and de-risk themselves. These and other networks also helped provide quality mentoring to the start-ups and entrepreneurs participating in PowerStart, the accelerator programme.

5.3.5. Building New Capabilities at CIIE

As mentioned, at the time of planning Infuse, CIIE lacked the capabilities required to raise commercial capital and invest it. Members of CIIE had managed seed funds, had significant experience of identifying and supporting promising start-ups

---

* For example, one of the companies, Altizon, was able to raise capital from Wipro, a leading Indian conglomerate. Similarly, Surya Power Magic raised money from Mahindra, another large Indian corporate, while Ecolibrium raised capital from JLL, which is a global real estate major.

through different programmes, and had built robust linkages with both corporate and non-corporate entities. The work done by the Centre in the entrepreneurship space and its links with IIMA gave it credibility and created confidence for potential partners. People with new skill sets were hired. An investment director with experience of working on energy policy at the World Bank and as a venture capitalist (including experience of investing in clean-tech space) filled a big void. In addition, two other persons, one with an energy-consulting background and the other with experience in strategic consulting, also joined the senior team. This senior team was supported by a couple of analysts. Moreover, the CEO of CIIE, who had been working with the government and other agencies in the clean energy space and who was instrumental in setting up Infuse Ventures, spent essentially all of his time overseeing the fund's activities. This Infuse team was complemented by a couple of venture partners with VC experience and by the investment committee which has members with experience of capital markets.

Building the Infuse team would not have been possible if CIIE did not have the flexibility that most incubators located in HEIs in India do not have. CIIE at IIMA was given the flexibility to hire for Infuse, the right people at market-linked compensation structures quite distinct from the pay scales and hiring policies of a government-linked academic institution. In fact, this flexibility was an extension of the autonomy that CIIE already had as an incubator-accelerator to follow its own hiring practices. This freedom, combined with the fact that CIIE was able to attract highly accomplished entrepreneurially minded technocrats and professionals, enabled it to put together a team of high achievers motivated by the challenge of creation rather than merely advancing their careers.

The team not only managed the existing investors but also others who helped in supporting start-ups and were also potential partners. Selling the concept of Infuse to potential partners/investors and helping them understand the value of portfolio companies was the single most important skill that the team leveraged. They always had their sales hat on which helped them get and retain partners! Despite several challenges, CIIE managed to deal with the stakeholders reasonably well and there were no fallouts. In fact, some of them continue to engage with CIIE for their new fund.*

Broadly then, what emerges from the discussion is that CIIE was a useful anchor for Infuse Ventures despite several constraints identified above, because they were able to build relevant skills reasonably quickly and manage stakeholders in an appropriate manner, although it was not very easy. Being part of a well-known academic institution, they were able to provide credibility for external stakeholders, both state and non-state ones. Besides, their earlier experience helped bring incubation and acceleration expertise to the table that the corporate sector did not have, and they were able to work relatively efficiently with the government. All this was critical to facilitate ecosystem building.

5.4. Strategy and Structure of Infuse Venture Fund

One of the elements that was critical to align the interests of different stakeholders was the structure of the fund and

---

* There have been instances wherein CIIE decided not to continue with some of the old partners in the new fund due to very onerous and bureaucracy-driven reporting requirements even though they were interested. In some cases, old partners were not very relevant any more as the mandate of the new fund was broader and different and did not match with the mandate of these partners.

its investment strategy and process. What fund structure is appropriate for initiatives like Infuse? What should the profit-sharing model be? How would the risk of failure be spread across different stakeholders? Who should make the investment decisions? What stage of funding should the fund be engaged in? These were some of the questions that needed to be resolved.

5.4.1. Investment Strategy

Given the broad mandate of ecosystem building and creating a pipeline for investments at the growth stage, Infuse invested at different stages through three pathways:

1. **Creation:** Infuse created ventures from scratch where necessary if unaddressed opportunities presented themselves. The investment adviser worked with entrepreneurs and industry partners to identify such opportunities and the right teams to take them forward.*
2. **Early-stage investment:** For very early-stage companies with potential, Infuse provided up to INR 5 million (approx. $80,000) of initial support with the potential for going up to INR 30 million (approx. $4,50,000) to pilot promising products and services.
3. **Scale-up and commercialization:** For companies that had completed a successful product/service pilot and had potential, Infuse provided up to INR 100 million (approx. $1.6 million) to support commercialization and the initial scale-up of the business. This facility was

---

* For example, Rajat Gupta engaged with Infuse as an entrepreneur-in-residence and co-created Tessol. Similarly, Infuse co-created cKers Finance from scratch in partnership with cKinetics.

available to companies in which Infuse had made early-stage investment as well as others.

The fund investment strategy was to essentially back start-ups whose products/services had a market potential, the associated technology risk was appropriate (i.e., not very high) and which had a good team in place. Infuse looked for high-quality, experienced and aggressive entrepreneurs with broad networks and then incentivized them with an ownership stake and options. The fund focused on businesses that addressed large problems with clear markets for the technology solutions. While Infuse invested in technology solutions, they did not support pure research and development. Investment was done to support companies that would commercialize existing technologies or adapt them to the Indian context. Even when start-ups were being created from scratch, solutions were being sought through combinations of existing technologies. In this way, Infuse sought to avoid taking on significant technology risk while still capturing the benefits of technology-focused solutions.

The business model of Infuse evolved over time but broadly, given the CoE mandate and the nascent nature of the clean-tech ecosystem, except probably in the beginning, a typical VC model was not adopted. While seed funding and acceleration support existed as part of the CoE mandate as well as part of CIIE's earlier work, the senior management group at Infuse realized that a grant pot needed to be created as many of the entrepreneurs did not really require $1,00,000 that was given, but often needed $10,000 to test their specific offerings. That was introduced in the middle of the programme to some advantage. Infuse started with a bunch of sector-agnostic accelerator programmes within the clean-tech space wherein start-ups with a focus on energy efficiency, renewables, water, sustainable agriculture, etc., all came together in one programme. Subsequently, it was realized that sector-specific programmes would be more efficacious as they would help organizers focus on a particular segment, and bring in the most relevant stakeholders for networking, mentoring and other support. As a result, programmes were organized around clean web, sustainable agriculture, water, etc. This focused approach provided more value to the start-ups as they were able to interact with stakeholders who were immediately relevant for them.

The other shift in the investment strategy was to increase the amounts invested through Infuse in early-stage start-ups from roughly \$2,00,000–\$3,00,000 to \$5,00,000–\$7,00,000 as start-ups needed more capital in this sector. Moreover, the composition of capital deployed also changed over time, especially in certain segments within the clean-tech space. It was realized that many of the start-ups required debt capital, especially those engaged in distributed renewable energy and focused on the rural segment. Fortunately, Rockefeller Foundation came forward as a part of their SPEED (Smart Power for Environmentally and Economically Sound Development) project to partner with CIIE and create a catalytic capital intervention through which debt capital was provided to set up mini and micro-grids across rural India. The spillover benefit of this intervention was that the senior team of Infuse realized that such capital is a big need for the start-ups. They subsequently joined hands with one of the consulting partners to create a non-banking financial corporation to provide debt capital. This company, CKers Finance, was supported by and invested in through Infuse Ventures.

5.4.2. Structure and Management of the Fund

In principle, structuring of the fund needed to be appropriate to take care of various stakeholder needs and constraints.

If the fund was not structured appropriately to make it incentive-compatible, its efficacy would be adversely affected. The fund's management consists of four principal parties (see Figure 5.1):

1. The Roman, *Investment Adviser* is the company that manages the fund, sources investments, undertakes due diligence and manages the portfolio. This role was performed by CIIE.
2. The *Trustee* represents the interests of the investors in the fund and has the responsibility for appointing the Investment Adviser, Investor Advisory Committee and Investment Committee.
3. The *Investment Committee* consists of the core team of Infuse and two board representatives of CIIE. The committee approves any investment and divestment decisions. Due to the fiduciary responsibility involved for the committee members, no investor representative is present on the committee. This also provides more autonomy to CIIE that is managing the fund.
4. The *Investor Advisory Committee (IAC)* is appointed by the trustee at the recommendation of the investment adviser to provide advice and counsel to the investment adviser, and help take some strategic decisions on behalf of the investors. All contributors, except BoI and UBI, had representatives on this committee. It was decided to give representation to large and strategic investors.

In the beginning of the fund cycle, CIIE (the investment adviser) also had the responsibility of raising funds to fulfil the matching grant obligation of the MNRE. IAC meetings and reporting were done on a quarterly basis. Updates on new investors and portfolio companies were done through these interactions. All conflict resolution was also the mandate of the IAC.

Infuse was largely structured as a regular VC fund but modified slightly to take care of the needs of various stakeholders. In a VC fund, the major issue is the alignment of interest between the investors and the investment adviser's team. Typically, in a VC fund, return proceeds (principal and profits) first get distributed to the investors till they have gotten a certain IRR (10 per cent hurdle rate) on their entire investment at the fund level. Thereafter, return proceeds get distributed between investors and investment adviser/team in the ratio of 80:20. This incentivizes the investment adviser to maximize returns. The innovation in Infuse was that the MNRE grant invested by CIIE was subordinated to the contributions from other financial/strategic investors. This meant that MNRE investment provided a risk cushion to the other investors—i.e., MNRE would recover their principal only after all other investors had recovered their principal and hurdle rate. This risk-cushion provision is similar to 'viability gap funding' or 'first-loss' protection, often provided by the government in a PPP project, and reduces the risk for investors if the project turns out to be unviable. This mechanism helped CIIE to 'crowd-in' more private capital as it reduced risk to investors.

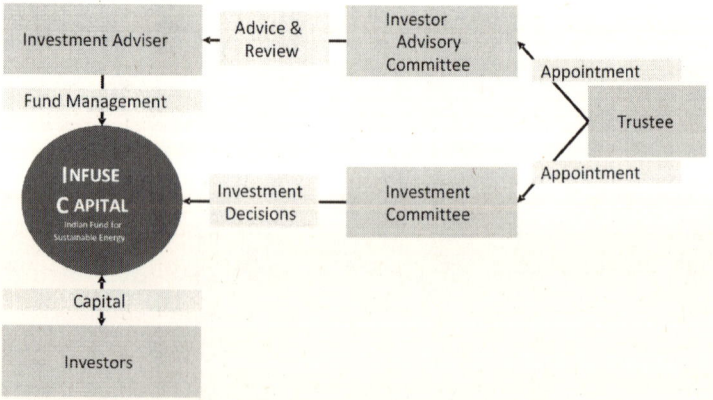

Figure 5.1: Fund Structure

Two other major conflicts of interest were envisaged. What happens if certain investors want to *directly* invest in a portfolio company of the fund? And, what if Infuse wants to invest in an incubatee company of the incubator, CIIE (where the investment adviser already holds a stake)? Such conflicts were resolved through the IAC.

Another potential worry that the investors, especially those representing commercial and strategic capital, could have had was that their contributions were being used not for investing in appropriate start-ups but for ecosystem-building. This was avoided because CoE undertook much of the ecosystem development work through the other initiatives, the resources for which were raised separately. Of course, the fund invested at much earlier stages of the start-up life cycle, took much riskier bets and invested in many more deals to accomplish the ecosystem objectives. If it was a regular fund, it may have invested in safer ventures and fewer deals.

The idea of blended capital, i.e., subordinating MNRE capital and using TA money to de-risk start-ups is very important for all developmental sectors. For example, the World Bank is trying to set up a fund in the agriculture space in Assam, and has learnt from the Infuse experience of blending different kinds of capital. Similarly, ideas relating to venture debt, autonomy of the investment adviser and the importance of raising 'acceleration' funds also have wider application in similar contexts. Finally, the opportunities of co-investment in portfolio start-ups that Infuse created and exploited is another dimension that is likely to be very useful in situations with high market, technological and regulatory uncertainties as it reduces risks for all stakeholders and can also benefit from economies of scale in funding start-ups. We shall return to some of these issues in the concluding section.

5.5. Some Concluding Observations

The investment period of the fund was five years. Almost all the capital raised for Infuse Ventures was invested by 2018 and other start-up support activities and objectives as set out under the MNRE-supported CoE have also been completed. Commitments on capital that remained un-invested in five years would lapse, as per a condition imposed by one of the investors (a standard feature in most VC funds). In hindsight, given our mission to address the market failure in early-stage clean-tech venture funding, this was a shortcoming of the fund structure. We had to allow a part of the capital commitments to lapse on this account. If Infuse had had the ability to continue investing in follow-on rounds, without the five-year investment period limitation, it could have utilized all the capital and start-ups would have benefited from the same.

A total of fifteen start-ups at different life cycle stages were funded by Infuse Ventures and over fourteen start-ups were provided seed grants while eight received debt capital. Over nine start-ups were provided acceleration support through various interventions by CIIE during this period. The companies supported were quite diverse and belonged to different segments of the clean-tech sector (see Annexure 5.1).

As expected in a sector like clean-tech, the mortality rate of the start-ups during the early stages is high. However, the companies supported by the CoE and its initiatives have, as of 2018, raised $27 against every dollar invested by MNRE, allowing it to overachieve the objective of crowding-in capital into the clean-tech sector set for the CoE by over 2000 per cent (Annexure 5.2).

Further, many of the portfolio companies of Infuse Ventures have become prominent names within the specific categories they operate in and have also received follow-on capital and validation from global strategic partners like Samsung and JLL, and Indian corporates like Wipro, Mahindra and TVS (Annexure 5.3). Infuse enabled the creation of companies which attracted strategic investors to take a bet on them directly as well. Further, global financial investors like TPG Rise, responsAbility and Calcef have invested in some portfolio companies—while Indian VCs like Siana Capital, Lumis Partners and Ankur Capital have invested in others. As on date, Infuse has also returned 55 per cent (i.e. distributed/paid-in capital multiple of 0.55) of the capital to its investors through two partial exits—putting it broadly within the global performance benchmarks, which was 0.5 for the 2000–17 period.[*][†]

While the direct financial and non-financial impact of Infuse and its start-ups may require a separate study, Infuse does offer a lot of learnings for any future experiment to catalyse entrepreneurship in new sectors and geographies. Based on the analysis so far, one can summarize the advantages of Infuse Ventures—a PPAP in the clean-tech space which is fraught with a variety of market and government failures. The advantages varied for different stakeholders (see Table 5.4).

---

[*] https://www.cambridgeassociates.com/wp-content/uploads/2020/03/Cambridge-Associates-Clean-Tech-Company-Performance-Statistics-3Q19.pdf

[†] This does not mean that Infuse did not miss out on good investments. In hindsight, two missed opportunities specifically stand out: Amplus Solar and Cleanmax Solar. Today, both are among the largest rooftop solar project companies in India that own and operate solar assets for their clients. They went on to scale successfully by raising large rounds of growth and strategic capital—Amplus from I-Squared Capital and Petronas; Cleanmax from Warburg Pincus and UK Climate Investments. A course correction was made later when Infuse backed another business in the same space, Fourth Partners, which has subsequently done well.

Broadly, the intervention of Infuse Ventures seemed to create incentive compatibility for all the key stakeholders in the clean energy sector. Different segments of the private players benefit as the risks, uncertainties and transaction costs they face are reduced through this intervention apart from reducing the risks associated with a weak ecosystem. Significant benefits also accrue to the academic incubator if it is able to build appropriate capabilities required to anchor such an intervention.

In Section 2 it was noted that given the risk-return-time profile, the traditional VC model may not be the most appropriate for supporting clean energy innovation (Gaddy et al. 2016). It was argued that the sector requires a more diverse set of actors and innovation models and the government can support increased participation of these actors in the clean-tech sector by lowering the risk of investments in the sector. Effectively, Infuse Ventures is one such policy experiment wherein government resources have been leveraged to raise more capital and bring together multiple stakeholders. Clean-tech requires patient capital, and by combining developmental, strategic and commercial capital, Infuse reduces the risk of commercializing clean-tech innovation and, to some extent, reduces the time and capital constraints typically faced by the private sector in this space. In fact, it undertakes a variety of tasks that Gaddy et al. (2016) recommend, given the inappropriateness of the conventional VC model for the clean-tech sector.[*]

---

[*] They suggest that corporations could also boost prospects for commercializing new technologies by strategically investing in or acquiring clean-tech start-ups, as happens in the biomedical sector. To attract the corporates in the sector, the government must share the risk of commercializing innovation by funding

Since private sector funds are not flowing to the clean-tech sector and direct intervention by the state has not been very efficacious, the state has a variety of policy options available. It can:

i.   Redesign existing instruments to take care of design faults but keep it under state control and implement projects on its own;

ii. Create more incentives and leave it to the private sector; or

iii. Experiment with a PPP model.

Of course, these are not solutions, and one can think of various permutations and combinations that the state can choose from. Given the government and market failures and the inappropriateness of the traditional VC model discussed in Section 2, the Infuse experiment makes conceptual sense. The fact that it has worked reasonably well suggests that such policy experimentation should continue. Infuse is not a conventional PPP model but a PPAP model. If done well, the participation of an academic institution through an incubator can build credibility and bring in critical incubation and acceleration experience. Despite the challenges in operationalizing the framework, the benefits that can accrue to all the stakeholders and the advantages of synergies that can be reaped (Table 5.4) are large enough to scale up such experiments.

In countries like India, most academic institutions are not very research-intensive and the quality of incubators is also quite uneven. Besides, a good incubator may not be

---

R&D and support projects with industry that demonstrate new technologies at scale (Gaddy et al. 2016:03).

co-located with a research-intensive institution and vice versa. While it might add to the organizational complexity of the policy experiment, it may be useful to bring a research-intensive HEI into this experiment apart from an incubator from a good HEI, preferably with management capabilities. One can argue for the use of private incubators but their ability to manage multiplicity of objectives (especially those related to developmental impact) and diversity of stakeholders may be limited. Entities in higher education that are not primarily profit-oriented bring in the right perspective to combine public and private goals. For CIIE, managing diverse objectives and stakeholders was not easy and therefore adding another dimension may enhance the organizational complexity of the initiative. But it will also bring in technical expertise and can open a window for accessing university labs and research centres, thereby enlarging the supply chain of clean-tech-related science and technologies. And addition of such dimensions can significantly impact the efficacy of such initiatives by reaping the economies of scope of diverse activities that are involved in creating and commercializing technology solutions in high-tech, high-impact sectors.

Infuse also provides a model for incubators' growth, diversification and financial sustainability. The transition of CIIE from being essentially a grants-based incubator-accelerator to an incubator-accelerator cum VC, which is able to augment its finances through management fees and crowd-in grant capital in the process, is a significant one. This seems to be an appropriate model for contexts where financial markets for early-stage investments are nascent and the entrepreneurial ecosystem is weak. Interestingly, MIT recently launched the MIT Engine—a fund-centric incubator model. From prototyping grants to acceleration and venture capital, to taking socially impactful innovations to the market are the activities it plans to focus on; almost exactly the same mandate as that of Infuse.* Apart from others, two interesting differences between MIT Engine and Infuse are worth noting. One, the fund size of MIT Engine at $150 million is much larger wherein MIT itself has invested $25 million as a limited partner. The initial contribution for Infuse came from a government entity. Two, unlike CIIE at IIMA, MIT is mainly a technology education institution but located in a multi-disciplinary context of not only MIT's own management and other departments but also in the broader Boston area which boasts of a vibrant entrepreneurial ecosystem. Finally, it may be useful to draw some general lessons for sectors like clean-tech where regulatory, market and technological uncertainties are high, resulting in situations where private capital may shy away and the state may not be a good substitute as they are not experienced technology investors and for other reasons discussed above. Obviously, patient capital is required in such situations. In fact, Gaddy et al. (2016) suggest that money could come from institutional investors like pension funds, sovereign wealth funds and family offices, which are not keen to realize returns very quickly. But experience elsewhere and the Infuse experience suggest that we also need other stakeholders to complement each other's capabilities and contributions.

The Fund-of-Funds Scheme (FFS) launched by the Department of Industrial Policy and Promotion (DIPP) and managed by SIDBI to back other third-party venture funds can be seen as similar to the contribution made by MNRE to CIIE to back Infuse Ventures. FFS is expected to reduce the risk of private capital and perform some of the functions that Infuse set out to do. There are significant differences between

---

* For details, see https://www.engine.xyz/

FFS and Infuse but certain lessons from Infuse may be useful and may help catalyse the flow of capital to start-ups as it was originally envisaged. The major challenge has been that VC funds have taken a long time to achieve their first closing and start investing. Based on the Infuse experience and other Indian and global best practices, minor tweaks to FFS can help unlock significant capital in India, especially in deep-tech and related areas which have characteristics common with clean-tech—where there may be market failures and commercial capital may be hard to attract easily.

Table 5.5 summarizes these suggestions which are self-explanatory. These are essentially focused on identifying areas in which the traditional VC model needs to be modified to make FFS more efficacious. Most of these tweaks have already been discussed in the context of Infuse and do not require repetition. The role that can be played by FFS as anchor but subordinated investor—standardization of documents, provision of draw-down and co-investment facility, importance of arms-length implementation to maintain autonomy of the investment adviser, enabling the use of CSR funds, availability of venture debt and acceleration funds—are all suggestions that are useful for FFS as they have proved to be effective in the context of Infuse and other similar contexts (see Table 5.5 for details).

Infuse Ventures is a one-of-its-kind policy experiment in which various stakeholders from the public, private and academic domains have come together to address issues faced by a technology-intensive, socially relevant sector wherein market and government failures are pervasive. While the jury is still out on the performance of the portfolio companies, the policy insights seem significant providing some pointers for the way ahead. To conclude, a few additional learnings may be highlighted and a few already mentioned reinforced.

Given the experience, if one wants to revisit the fund design and its mission, the ideal intervention would probably have been a long-dated multi-stage dual-mission clean-tech fund. To elaborate, Infuse's mission was to address the 'market failure' in early-stage funding for clean-tech ventures, but the reality was that the 'market failure' existed across stages—early-stage financing, growth capital, debt and strategic investments as well as M&A. The implicit hope was that these gaps in subsequent stages of the ecosystem would be addressed by markets in due course, especially since ventures coming out of Infuse's portfolio would have a lower risk profile (as early-stage risks were assumed by Infuse at the stage it invested in these ventures). But that did not come to pass. Commercial and strategic growth capital for clean-tech continues to be hard to come by. Consequently, it has not been easy for Infuse to raise follow-on growth capital funding and create exit options in its portfolio. While CIIE as fund managers recognized the gap in availability of debt finance and co-created cKers, the clean-tech debt financing venture in partnership with cKinetics, it did not take on the challenge of setting up a parallel Infuse Growth Capital Fund (to address the gap in growth equity financing). This was not without a reason. A larger growth fund, with a private equity model, would have required a whole new team and competencies that CIIE did not have the bandwidth for and it would have meant an even more dramatic transition for an incubator-accelerator.

Since the success of interventions in early-stage clean-tech finance is dependent on the availability of growth equity, one cannot solve the former problem without addressing the latter. So, Infuse should probably have had a dual mission of early-stage and growth capital intervention. Of course, to facilitate that, the fund requirements would have been much larger, with a mandate to invest in both stages—early and growth. Such a mandate would necessitate a long-dated fund, with a duration of fifteen–twenty years (instead of the usual eight–ten years), allowing the provision of a line-of-equity to Infuse portfolio ventures (linked to performance) that extended support from early to growth stages. The key question is, was raising such a fund even feasible at that stage? The model envisaged in the experiment was so novel to begin with, a more ambitious version may not have found support from institutions providing development, commercial and strategic capital. Moreover, the challenges to attract the right kind of team to manage such an early-stage/growth capital fund would have been even more difficult to overcome. Given this context, Infuse Ventures was a very useful experiment that has provided important insights for future policy interventions.

Table 5.4

Infuse Ventures and Potential Benefits for Various Categories of Stakeholders

Stakeholder	*Potential Benefits for the Stakeholder*
Government	• Reduces the risk of government failure in choosing inappropriate projects • Gets access to better investment and management expertise • Improved ability to leverage financial capital and therefore higher impact • Engages closely with entrepreneurs and innovators to think through new ideas within the sector • Supplements the grant-based R&D efforts by the government with market-linked investing • Useful policy inputs from the ground
Developmental Capital	• Financial contribution leveraged better • De-risked investment while still allowing achievement of the developmental goal • Lower transaction costs of choosing and managing projects • Access to capabilities and resources of other stakeholders • Reduction in market, technological and regulatory uncertainty
Strategic Capital	• Lowering of business and strategic risks • Financial contribution leveraged better • Lower transaction costs of choosing and managing projects • Access to capabilities and resources of other stakeholders • Easy access to market insights w.r.t. new innovations and business models leading to 'fast second strategy' ability to invest or acquire • Reduction in market, technological and regulatory uncertainty

Commercial Capital	• Lowering of business risks through blending with strategic and development capital • Financial contribution leveraged better • Lower transaction costs of choosing and managing projects • Access to capabilities and resources of other stakeholders • Reduction in market, technological and regulatory uncertainty
Other Grant-Making Partners	• Contribution leveraged better due to economies of scope in related activities • Lower transaction costs of choosing and managing projects • Access to capabilities and resources of other stakeholders
Academic Incubator	• Enhanced scale and scope of operations with potential of reaping benefits of the same • Improves ability to provide better incubation or acceleration support • Aids financial sustainability of the incubator • Aids building and strengthening of networks

Table 5.5
Implementing Fund-of-Funds Scheme: Some Learnings from the Infuse Experiment

Area	Recommendation	How will it help?	Evidence from Infuse Experience and Elsewhere
How can FFS help VC funds complete fund-raise and start investing faster (vs the present twelve- to twenty-four-month cycle)?	1. Create a **warehousing facility** under FFS to enable VC fund to start investing while they fundraise.	Approved fund managers will be able to start co-investing in deals immediately upon FFS approval vs waiting for months to start investing. This will also help build track record of fund managers and attract capital.	This is a common provision in funds but seldom used, as fund manager may not have funds. CIIE had used MNRE and other available funds to make seed investments and warehoused investment prior to initial closing of Infuse. Helped get six–twelve months' head start while fundraising efforts for Infuse continued.
	2. FFS to act as an **anchor investor** especially in tough technologies areas by committing large and committing early.	Being an anchor investor by being the first limited partner (LP) to commit, and making government commitment as high as 50 per cent to the fund corpus will help achieve the fund-raise faster.	MNRE's upfront commitment of 50 per cent (i.e. mandate to raise only matching amount) helped CIIE exceed the matching dollar target from other investors, and get over 4x fund commitment. Elsewhere, Yozma scheme in Israel matches dollar to dollar. Singapore government's SPRINT scheme matches dollar to dollar.

	3. Standardize Indian **fund documents and terms** through global benchmarking.	Currently, significant time is lost in bringing all domestic investors to common terms. Several FFS terms are also not in line with global best practices. This will reduce negotiation and make all domestic investors adopt the FFS standard.	CIIE lost over a year negotiating documents with initial investors like MNRE, BP and TDB—but all documents got renegotiated when global investors like IFC joined the fund, and brought in some international best practices. Globally, accelerators like YCombinator had created standard terms and open-sourced documents for use by other investors for faster closure.
How can FFS catalyse/ crowd-in more capital for VC funds?	4. FFS to act as a **subordinated investor** in VC funds focused on tough technologies.	FFS providing part of its commitment as subordinate to commercial LPs investment will reduce the downside risk for the LPs.	In Infuse, MNRE's capital was subordinated to the commercial investors and helped attract unexpected sources of capital by providing an additional risk cushion. Viability gap funding is common in PPP projects. Liquidation preference for new investors is a common provision in equity investments to attract more capital.

	5. FFS to provide **buy-back option** to fund manager and other LPs.	Fund managers and LPs will get the option to maximize their upside by buying the FFS stake in the fund at the end of the investing stage.	Yozma and Singapore government models had a provision to buy back the government stake after charging minimal interest This was considered but not adopted by Infuse since downside risk coverage through subordinated capital was more critical.
How can FFS catalyse more early-stage investing?	6. FFS to back creation of small **accelerator funds** (of ~ INR 100 crore fund corpus) in India, initially at existing academic incubators, by committing up to 65 per cent.	Accelerator funds have a high throughput as they invest in batches and provide initial seed funding (up to INR 50 lakh per start-up) to several start-ups in one go. This will get VCs and angels to actively engage in the selection of start-ups.	Through accelerator-driven investment process through programmes like PowerStart, RESeed and Financial Inclusion Lab, CIIE has been able to make over 100 investments and attract the right talent. YCombinator and SOSV are great examples of the accelerator VC model. In the academic world, StartX at Stanford and Engine at MIT are good examples.

| | 7. FFS to create a **co-investment facility** to co-invest along with other VC funds, angels and corporates. | Co-investment along with other investors but with subordinated liquidation preference will enable more capital for the start-ups—and somewhat reduce the risk for the co-investor. This will encourage investors to take greater risks with early-stage investing. | Singapore government's SPRING initiative has an extensive co-investment model to co-invest along with VC funds, angel investors and corporates— wherein it provides matching investment to the extent of $2 million per start-up. Infuse also used this model but with limited success. |
| How can FFS remove uncertainties and bureaucracy? | 8. The government should continue its policy of keeping FFS at arm's length from implementation to ensure efficiency and to maximize continuity. | Government involvement in implementation is likely to hamper FFS's impact—as government is not geared to directly invest in funds or start-ups. | Infuse fund structure provided significant autonomy to CIIE. However, frequent bureaucratic changes within the government made continuity difficult—and impacted the fund adversely. *Boulevard of Broken Dreams* by Josh Lerner is full of examples where direct implementation or over-involvement by government has been catastrophic. |

| | 9. FFS must create satisfactory **drawdown provisions** to ensure government does not default. | Creating a very clear mechanism for flow of funds from DIPP to SIDBI shall ensure that no default happens in terms of honouring an investment in a VC fund, when funds are drawn-down. While the other investors in a VC fund may be used to contributing their capital at a ten to fifteen days' notice the government may not be geared up to do the same. | In Infuse, on a few occasions, government departments like MNRE and TDB made significant delays in making draw-down on time, amid changing leadership and officials. But, in principle, Infuse had this facility which ensured that CoE had sufficient capital from MNRE to contribute towards Infuse whenever required. Annual or bi-annual lump sum capital calls from the government makes more sense. |
| Beyond FFS, what can the government do to catalyse investments? | 10. Give corporates and individuals **tax deduction benefits** upon investing in start-ups or VC funds. | Bring investing in start-ups in line with tax benefits for investing in R&D. | The Singapore government offers 50 per cent tax deduction to investments made in start-ups.

R&D investments in India have 150 per cent tax deduction. |

	11. Enable **CSR and foundations** to invest in social venture funds.	Current CSR regulations and income tax regulations do not permit CSR and foundations to invest their capital in social venture funds or start-ups. This will unlock a new form of capital for the VC funds.	In the US, foundations are allowed to make Programme-Related Investing (PRI) in impact-focused initiatives including VC funds and start-ups that deliver social benefits. Infuse was unable to raise such capital from Indian foundations. However, in subsequent funds, CIIE has been able to attract global foundations like Dell Foundation, Soros Foundation, among others.

	12. Enable **venture debt** to become available more easily to start-ups.	The risk guarantee scheme for start-ups awaiting Cabinet approval shall be a great move as start-ups will not have to depend on expensive equity for scaling up.	Start-ups require significant debt as working capital to scale up. Infuse start-ups significantly benefited from debt provided by banks under the Credit Guarantee Fund Trust for Micro and Small Enterprises (CGTMSE) scheme of GoI. Infuse implementation also brought this out, leading to the creation of a start-up to provide debt capital. Seeing the need for debt, Infuse also created cKers Finance to support ventures in the sustainability space through debt which has been well received by the ecosystem. The venture debts model has worked extremely well across the world and reduced the equity requirements of start-ups.

Annexures

Annexure 5.1

Start-ups Supported by the CIIE CoE through Various Mechanisms and Their Current Status			
A. Start-ups Supported by the CIIE CoE Before Infuse Became Operational			
#	Name	Description	Current Status
1	Avani	Biomass (pine) supply chain and energy generation	Not operational
2	Ecozen	Solar-based agri pumping and cold storage solutions	Scaling up
3	Insolare and iGrenenergi	Micro-inverter for improving efficiency of solar modules; Energy Performance Contracting (EPC)	Scaling up
4	Boond	Rural solar products and EPC	Scaling up
5	Greenway Grameen	Efficient biomass cook stoves	Scaling up
6	Onergy	Rural solar lighting, mini-grids	Scaling up
7	Flybird	Solar-based smart irrigation controller	Scaling up
8	Ecolibrium	Smart Grid—energy management and RE integration	Scaling up
9	Banyan	Plastic recycling and processing solutions	Scaling up

B. Start-ups Invested in by Infuse Ventures

10	Ecolibrium	Smart Grid - Energy Management and RE Integration	Scaling up
11	Fourth Partner	Solar EPC and financing platform	Scaling up
12	GIBSS	Geo-cooling for industrial and commercial applications	Shut down
13	cKers Finance	Financing platform for early-stage renewables start-ups	Scaling up
14	Altizon	IoT-based solar and equipment monitoring solution	Scaling up
15	Proklean	Non-polluting industrial chemicals	Scaling up
16	Karma	Electronic waste recovery and recycling	Shut down
17	Silvan	Home energy management system	Scaling up
18	Visviva	Biomass supply chain	Shut down
19	Surya Power Magic	Solar-based agri pumping and farm productivity solutions	Shut down
20	REConnect	Renewable energy services including renewable energy credits (RECs), wind forecasting	Scaling up
21	TESSOL	Electric cooling reefer vehicles	Scaling up
22	Ezysolare	Online platform for streamlining purchase of solar components, tender	Pivoted to other areas
23	Glowship	Online platform for purchase of environment and solar solutions	Slow progress
24	Revive	Waste pyrolysis-based e-waste recovery process	Slow progress

C. Start-ups Supported by CoE through RESeed			
25	Promethean Energy	Solar-thermal and waste heat recovery for industrial applications	Scaling up
26	RenxSol	Solar thermal solutions	Scaling up
27	Solarwaale	Online platform for solar pipeline	Not operational
28	Sunkalp	Online platform for solar lead generation	Not operational
29	New LeafDynamics	Biomass-based cold chain system	Scaling up
30	AlgoEngines	IoT-based cloud monitoring solution for solar and wind	Scaling up
31	Ronds	Battery management solution for solar installations	Scaling up
32	Brisil	Conversion of rice-husk biomass into valuable products	Slow momentum
33	Cloverboard	Home energy management system	Not operational
34	E-Cube	Industrial energy management solutions	Scaling up
35	Clytics	Smart grid and energy disaggregation platform	Slow momentum
36	Energyly	Consumer energy consumption application	Slow momentum
37	Pirhoalpha	Building management system	Scaling up
38	OpenWater	Energy-efficient water purification system	Scaling up

D. Start-ups Supported by CoE through Catalytic Capital Facility Created with cKinetics and through cKers Finance

39	Freespanz	Solar-based mini grids	Scaling up
40	OMC	Solar-based mini grids	Scaling up
41	DesiPower	Biomass-based mini grids	Scaling up
42	Punam Energy	Rural mini grids	Scaling up
43	Sure Energy	Rooftop EPC	Scaling up
44	Claro	Rural water pumping	Scaling up
45	Sun Source	Rooftop EPC	Scaling up
46	SunSure	Rooftop EPC	Scaling up

Annexure 5.2

Sectoral Profile of Investments, Ecosystem Development KPIs and Other Achievements (As of 2018)

Key Performance Indicator	Target	Achieved
No. of aspiring entrepreneurs trained	100	200
Total no. of clean-tech start-ups supported	20	40
Renewable start-ups supported	20	26
Other clean-tech start-ups supported	-	14
Funds channelized in RE Start-ups vs Planned (in crore)	39	59

Types of Clean-Tech Sectors Supported	No. of Projects
Solar	17
Biomass	6
Energy Efficiency	6
Waste	3
Wind	2
Geo-cooling	1
Hybrid Vehicle	1
Water	1
Air Quality	1
Green Chemicals	1
Smart Grid	1

Summary of Some Key Achievements of Infuse Ventures	
Leverage on MNRE's funds vs planned (1x)	27x
More funds generated from non-MNRE sources vs planned	23.5x
More funds channelled for start-ups vs planned	17.3x
More funds channelled for RE start-ups vs planned	15x
Achievement on no. of aspiring entrepreneurs trained	200 per cent +
Achievement on no. of start-ups supported	200 per cent
Achievement on no. of RE start-ups scaling up vs planned (15)	133 per cent
Percentage of generated funds used for RE vs planned (50 per cent)	88 per cent

Annexure 5.3

Infuse Supported Start-ups that Received Follow-on Strategic and Financial Investments	
Company	Investors
Ecolibrium	JLL
Fourth Partner	TPG Rise
GIBSS	responsAbility
cKers Finance	Calcef
Altizon	Lumis Partner, TVS, Wipro
Proklean	Siana Capital
Silvan	Samsung
Surya Power Magic	Mahindra
Tessol	Ankur Capital

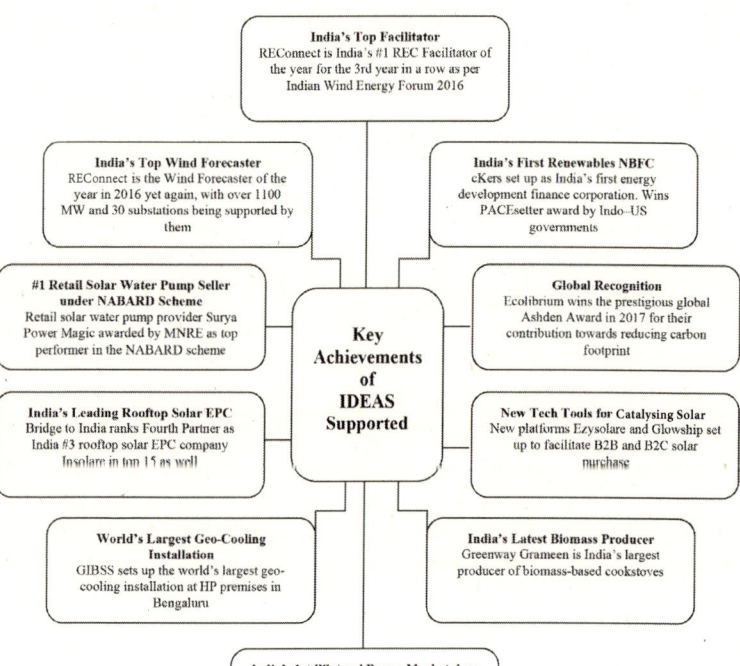

India's Top Facilitator
REConnect is India's #1 REC Facilitator of the year for the 3rd year in a row as per Indian Wind Energy Forum 2016

India's Top Wind Forecaster
REConnect is the Wind Forecaster of the year in 2016 yet again, with over 1100 MW and 30 substations being supported by them

India's First Renewables NBFC
cKers set up as India's first energy development finance corporation. Wins PACEsetter award by Indo–US governments

#1 Retail Solar Water Pump Seller under NABARD Scheme
Retail solar water pump provider Surya Power Magic awarded by MNRE as top performer in the NABARD scheme

Key Achievements of IDEAS Supported

Global Recognition
Ecolibrium wins the prestigious global Ashden Award in 2017 for their contribution towards reducing carbon footprint

India's Leading Rooftop Solar EPC
Bridge to India ranks Fourth Partner as India #3 rooftop solar EPC company Insolare in top 15 as well

New Tech Tools for Catalysing Solar
New platforms Ezysolare and Glowship set up to facilitate B2B and B2C solar purchase

World's Largest Geo-Cooling Installation
GIBSS sets up the world's largest geo-cooling installation at HP premises in Bengaluru

India's Latest Biomass Producer
Greenway Grameen is India's largest producer of biomass-based cookstoves

India's 1st Bilateral Power Marketplace
India's first bilateral renewable energy and power platform Clickpower developed and waiting CERC approval

6

In Conclusion

Exploring the Way Ahead

6.1. Introduction

The objective of innovation policy is to positively affect innovation activities of various entities in an economy and their innovation outcomes. Innovation is desirable as it enhances productivity and facilitates economic growth. The book provides a broad framework wherein the concept of innovation, the nature of innovation-related activities and the understanding of the innovation ecosystem are integrated in order to understand the complexity and nuances of the innovation process. This task is undertaken with the expectation that such an integration would help analyse various dimensions of innovation policy and its likely impact on innovation efforts and outcomes in a meaningful manner.

At the core of the book is the simple argument that an effective innovation policy requires an appropriate conceptualization of innovation and understanding of innovation activities. Else, any analysis of the policy instruments that can potentially affect the nature and extent of innovative activity in an economy is likely to be incomplete. It is posited that technological innovation can broadly take three forms—*products*, *processes* and *practices* (*3 Ps*). While innovations can happen independently in the three domains, the 3 Ps are often linked, with innovations in each P feeding into innovation activities in the others. New processes in the form of new machines can facilitate the development of new products and new practices can enhance the efficacy of product and process innovations. The novelty of the innovation can be defined at the firm and the market (both local and global) levels. Innovations that are new to the firm but not to the market are critical for an economy as they reflect diffusion of innovations which affect economy-wide productivity growth. Therefore, an exclusive focus of policy on innovations that are new to the market—local, national or global—is misplaced. The trade-offs between innovation and diffusion are real and need to be addressed and managed.

Innovation activities are varied and not always easy to capture empirically. Broadly, these include all those activities that help develop or source knowledge relevant for innovation. R&D activity, technology purchase and efforts to learn from innovations of others are essential parts of these endeavours. R&D can be done in-house, outsourced or in collaboration with others. New technology can be purchased in an embodied form (e.g., a new machine) or in a disembodied form through licensing, etc. Learning from others or exploitation of knowledge spillovers can also take various forms including reverse engineering, hiring of consultants or employees of other firms and so on. Collaborations for R&D can be global or local, new technology purchases can be domestic or foreign, and knowledge spillovers can be regional, national and global.

Business enterprises (both public and private), the government at different levels and academia (specially

HEIs)—the three key pillars of the NIS—are important participants in innovation activities. Meaningful linkages and interactions among these entities are important for generating innovations. But these interactions are not, and should not be seen as, uni-linear or uni-directional, but iterative and reciprocal. Only when reciprocal synergies among various elements of the NIS are reaped, can one expect to develop a vibrant innovation ecosystem. Given significant developments in science and technology along with convergence across various technologies, the innovation process has become very complex, enhancing the need for continuous learning and absorption capabilities for competitiveness at all levels of an organization. Consequently, the complexity of interactions across stakeholders has increased along with the importance of such interactions for learning. Given that the interaction between different entities within the NIS is critical for innovation outcomes, the success of innovation policy instruments is likely to depend on how policy is able to encourage and facilitate the flow of knowledge across the NIS of an economy and also improve the chances of it being combined and implemented in innovative ways. Studies show that the possibility of combining ideas from different sources improves significantly if organizations have capacities to receive and use the knowledge embedded in these ideas. Building of such innovation capacities or capabilities has to be an important focus of policy.

Once these features of innovation and innovation activities are recognized, the scope of innovation policy becomes broader and analysis of the innovation-public policy interface becomes even more complex. Earlier chapters have analysed a large body of theoretical and empirical literature to explore the contours of innovation and public policy linkages and highlighted their complexity. Innovation policy in this analysis is defined very broadly to include all policy instruments that affect the choices of economic agents with respect to innovation efforts and outcomes. In order to understand the current policy challenges facing India and the way forward, an effort is made to analyse recent trends in innovation and innovation activities in the country. This is done in the context of studies undertaken in various parts of the world, including India. While the analysis covered a variety of policy instruments which can influence innovation, recent policy initiatives to support innovation-driven entrepreneurship is one of the key focus areas of the analysis undertaken in the book. This chapter makes an effort to put together various strands in the discussion so far to identify a few areas which require early attention of the Indian policymakers.

Broadly, it is argued that policy formulation needs to consider complementarities across various instruments of innovation policy and focus sharply on those instruments that have the maximum externalities for other policy initiatives. Given this perspective, we highlight below some key areas that need immediate policy attention.

6.2. Innovation Policy in India: Some General Issues

It is evident that the theoretical and empirical studies on the impact of different policies on innovation are mixed and somewhat difficult to interpret, especially due to the conceptual complexity of defining innovation and issues relating to the measurement of innovation and policy instruments. Can one then draw any conclusions about the efficacy of different policy instruments and identify salient areas of intervention for innovation-driven long-term productivity growth in India?

Despite all the analytical complexities, one of the most consistent results in the available literature is that *all* policies that can potentially affect innovation-related outcomes—R&D, technology licensing, IPRs, industry, trade, FDI—tend to have a positive impact in situations where innovation capability and absorptive capacity is reasonably high. Also, innovation policy instruments are more efficacious if levels of competition and openness are high. In other words, policies like an appropriate IP regime, tax and other support for R&D, when combined with liberalization on various fronts (trade, FDI, technology licensing, industrial investment) that enhances competition and contagion potential, are likely to positively influence innovation outcomes in the presence of good innovation capability. Innovation policies, therefore, not only need to encourage innovation capacity-building but also ensure contestable markets.

These results would suggest that the success of any liberalization process will crucially depend on the quantum and nature of investments that domestic firms make in building their innovation or technological capabilities. The major policy challenge, therefore, is to identify instruments which induce such investments without curtailing the flow of technologies from abroad. The analysis contained in the book does not provide an unequivocal verdict on the impact of liberalization on innovation in India. Liberalization entailed significant FDI inflows, increase in trade flows, spurt in foreign collaborations, joint ventures and marketing tie-ups. The post-reforms years have seen an improvement in India in terms of its focus on conventional innovation inputs (R&D) and outputs (patents). The increase in the private sector share of Indian R&D is noteworthy but it continues to be much lower than other nations. Besides, despite an increased focus on R&D, its intensity (R&D to sales ratio) remains low for Indian firms. And while patenting activity has increased significantly as compared to the earlier situation, India remains far behind countries like China. Consequently, the changes in R&D and patenting are neither dramatic nor consistent in the post-liberalization period and India remains a laggard in a relative sense. Overall, Indian firms seem to remain somewhat reluctant to consistently invest in innovation activities to attain technological dynamism and competitiveness.

Foreign firms in India have also not shown a higher R&D orientation either and R&D intensity in such firms is similar to domestic firms. However, a large number of MNCs have set up R&D centres in India which are very active in patent filings. Overall, it appears that innovation-related activities of MNCs are yet to create significant learning potential for domestic firms. While it is difficult to measure the impact of MNCs' presence, it is more likely to be through *competition* rather than *contagion*.

At the aggregate national level, there are also some concerns about excessive outflow of resources for purchase of foreign disembodied technology through licensing. R&D expenditures have grown at a slower rate as compared to expenditures incurred on technology licensing. This tilts the ratio in favour of the latter, reflecting India's 'technology dependence'. However, at the firm level, one does not see a significant increase in the use of foreign technology licensing in recent years. But there was a fairly long phase (especially in the early 2000s) during the post-liberalization period, when embodied technology purchase from foreign sources grew very rapidly, although it has seen a decline in recent years. Adoption of new machinery (imported or otherwise) reflects process innovation at the firm level. This has also been combined with investments in ICT which are likely to have facilitated the adoption of new processes and practices. Although ICT expenditures have risen, their share in capital formation and value added has not grown significantly and remains low by international standards.

6.2.1. Contestability-Related Issues

While the jury is still out on the impact of extreme competition on innovation, most studies show higher innovation activity in contestable markets. In the post-liberalization period, contestability in Indian markets has been enhanced through a variety of policies whereby restrictions on entry, both for foreign and domestic entities, were removed. As discussed, the growth of total factor productivity (TFP) in Indian manufacturing grew during the post-reform period and was the highest during 2001–09 when competition through import of final goods was growing and intensity of use of imported embodied technology (capital and intermediate goods) was also rising rapidly. The intensity of use of disembodied foreign technology through licensing also grew during this period albeit less rapidly. However, R&D intensity did not show a consistently rising trend. So, an open economy seems to have helped the growth of TFP in the Indian manufacturing sector through a variety of innovation-related activities, of which R&D probably was not the prime one. Given the fact that typically R&D is a riskier investment than technology purchase, one would expect enterprises to resort to technology purchase (both embodied and disembodied) when faced with increase in competition, provided policy permits them to do so. Over time, one may see more reliance on R&D.

In recent years, one has seen a movement against trade liberalization, with several economies trying to close their markets to imports and move towards self-reliance. India is not an exception with *Atmanirbhar Bharat* becoming a clarion call. It is not clear what form this broad policy thrust would take as policymakers are yet to spell out a coherent policy strategy or framework.* It will be disastrous if the nation reverts to the License Raj or to an era of import substitution. Tinkering too much with the trade policy or reversing the trend of liberalization by pulling back other liberalization measures is unlikely to help innovation-based competitiveness and may actually result in government failures as different sectors would lobby for protection of their markets from imports and probably even local competition. Given the integration of Indian markets with global markets, such a move will not only distort firms' cost structures, it may also result in losing the efficiency gains that Indian industry has made through adoption of new technology and incremental innovations that are difficult to observe and document.

In the pre-reform era, the major objective of R&D probably was to facilitate policy-induced import substitution and the indigenization process. Competition pressures being low, cost and efficiency were not the primary considerations, with self-reliance and foreign exchange savings being uppermost in the minds of the policymakers. Import substitution policies induced adaptive R&D and developed specific kinds of technological capabilities. Firms used inward-oriented policies to garner large market shares in the production of

---

* A Make in India programme, which can potentially be seen as a part of the Atmanirbhar Bharat campaign, was launched by the Government of India in 2014 to encourage firms to manufacture their products in India. In order to attract manufacturing investment and transform India into a global manufacturing and design hub, an environment conducive to investments was to be created by developing modern and efficient infrastructure, and allowing FDI into new sectors. A variety of initiatives have been launched as a part of this programme, but the overall coherence of these initiatives and the extent of their impact remains unclear. For programme details, see https://www.makeinindia.com/

import substitutes. They took up one import substitute after another—moving on to the next one when the profits on the first one declined. This resulted in over-diversification and fragmented R&D, without the acquisition of technological mastery in any one field. With a policy shift towards autarky, the same kinds of processes may get rejuvenated and Indian industry may not be able to build capability to survive and grow in a competitive environment.

A movement towards autarky will also deprive Indian firms of learning from competing in demanding export markets. As discussed, most late industrializing nations of East Asia and China have built their innovation capabilities through export-led growth. In the context of China, many argue, this was made possible through an undervalued exchange rate.[*] Significant growth in exports provided scale advantages and the linkages with the external world facilitated building of innovation capabilities. Insofar as exchange rate policy can facilitate such opportunities, India should also consider an appropriate policy in this space. An undervalued exchange rate will spur exports but will make imports more expensive which will add to costs of firms that are import-dependent, making them less competitive. Besides, more expensive imports will also put a brake on the processes of learning emanating from imported machines and other inputs. The two processes of learning need to be balanced, which is not an easy task. It also needs to be recognized that an overvalued exchange rate combined with a high cost of capital, discussed in some detail below, can make export-led growth decidedly difficult. In any case, only exchange rate policy may not help Indian firms penetrate export markets and many other things, some of which are discussed below, need to be in place.

---

[*] See, for example, Morris (1997); Cardoso and Duarte (2017).

As a part of the Atmanirbhar Bharat agenda, the Government of India has introduced a Production-Linked Incentive (PLI) scheme to encourage indigenous high-tech manufacturing and reduce the import bill. The scheme, initially introduced for mobile phones and other electronic components, has recently been extended to other sectors like food processing, telecom and networking products, man-made fibre and technical textiles, specialty steel, automobiles and auto components, high-efficiency solar photovoltaic modules and white goods such as air conditioners and LEDs.* The policy is based on the premise that manufacturers in these sectors do not have a level playing field vis-à-vis competing firms in other nations. Indian firms suffer on account of high cost of finance, inadequacies in infrastructure (especially power, domestic supply chains and logistics), limited design capabilities and skills gaps. The scheme seeks to incentivize the setting up and expansion of new manufacturing facilities in these sectors, both by foreign and domestic firms. It is too early to assess the impact of this policy but if it results in the expansion of high-tech manufacturing in India by MNCs and domestic firms, both contagion and competition effects would result in more innovative efforts. However, it is not clear how the government would ensure consistency of this policy with the growing tendency towards autarky as trade barriers are being raised. Higher trade barriers may hamper the effective use of the PLI policy for those high-tech industries where manufacturing entails use of critical imported inputs.

All this does not mean that the state should work only with macro policies and not make direct efforts to help build local capabilities in specific sectors and technology

---

* For details, see https://pib.gov.in/PressReleasePage.aspx?PRID= 1671912 and https://www.meity.gov.in/esdm/pli

domains that are critical for technology dynamism in the future. But these efforts need to be thought through systematically to develop appropriate policy instruments, rather than using self-reliance as a political rhetoric with no coherent strategy. One possibility is to adopt some mission-oriented innovation policies. We will explore this option in the latter part of this chapter.

Broadly, in an open economy and contestable markets, interactions between foreign technology flows and indigenous efforts are crucial for building innovation capabilities. Foreign technology flows through; imports, exports, FDI and other linkages create opportunities to learn. But we know from our earlier discussion that policy choices with regard to foreign technology flows of any variety are more complex than many would like us to believe. Exploiting learning opportunities from the existence of foreign technology spillovers are affected by the quantum and nature of knowledge flows as well as the local capabilities to assimilate and build on those knowledge flows. Thus, while policy needs to encourage local technology efforts, it also should incentivize larger and better-quality knowledge flows into the economy. With such knowledge inflows, the learning is much higher if one is able to export to more demanding markets and graduate to more technology-intensive exports over time.

Learning is also facilitated if MNCs build linkages with local entities—enterprises as well as HEIs—for research and manufacturing. Mandating such an activity by MNCs may not work in the case of India although nations like China have successfully adopted such policy instruments. The demand and supply conditions in India are not so lucrative for MNCs as they are for the Chinese economy. In the case of China, both market and resource-seeking opportunities are higher, making MNCs more amenable to certain conditions. It can, of course, be argued that in the changing geopolitical situation with China somewhat on the back foot, India may have a greater degree of freedom in this respect. While available policy options to encourage foreign knowledge flows need to be exploited, some of the policies discussed below may provide the conditions that will make MNCs more prone to undertaking knowledge-intensive activities in India.

Apart from industry, trade and FDI policies, IPR-related regulations can also affect degrees of contestability in markets. A stringent IPRs policy, implemented well, can create entry barriers as owners of IP like patents can enjoy monopoly rights. As discussed in the earlier chapters, the relationship between IP and innovation is quite complex and there does not seem to be any evidence to show that innovativeness of firms in an economy consistently increases with higher degrees of IP protection. Some studies have alluded to a non-linear (inverted U) relationship between stringency of IP protection and innovation, implying that a moderate degree of IP protection is the most appropriate as it also helps the diffusion process. At the same time, other studies have highlighted that understanding the context is critical for fine-tuning IP policies for each country and homogenizing IP policies across the world does not necessarily make sense. The World Trade Organization agreements provide opportunities for member nations to be somewhat flexible in deciding their IPR policies. Countries like India have used these opportunities well.

6.2.2. Cost and Availability of Capital

The fact that the cost of capital in India has been high relative to other nations was highlighted earlier. It was also shown that the availability of early-stage financing for innovation-driven start-ups has also been low although one observes some positive trends in recent years. Therefore, just like all other investments, innovation financing is a key problem in the country. Apart from undertaking R&D on their own, buying technology and learning from innovations of others are important innovation-related activities. Insofar as adoption of external technology, both embodied and disembodied, is a less risky (vis-à-vis R&D) and quicker mechanism to deal with enhanced contestability, availability of capital at reasonable costs is also critical for adoption of external technology-driven innovations at the firm level. Moreover, commercialization of innovation often requires significant investments to access relevant complementary assets to take the innovation to the market. High costs of capital make such investments expensive, making innovation activity even more risky. That is, even if an enterprise is able to put together resources to develop an innovation and perform that task successfully, it can only create an appropriate value if it has access to capital at reasonable costs to invest in complementary assets and take the innovation to the market. The problem gets exacerbated if the required complementary assets for commercialization are specialized or specific to the innovation and not easily available within the enterprise or in the market.

These assets may take different forms like new manufacturing systems, marketing and sales networks, after-sales support, new skills, etc. Typically, investments in *specialized* assets require more capital than in *generic* assets (e.g., widely used manufacturing processes, marketing and after-sales support systems) that are already available. Moreover, investments in specialized complementary assets are risky because such investments become a sunk cost if innovation fails as the assets are specific to the innovation and cannot be redeployed easily for other purposes. Therefore, it is critical for policymakers to recognize that it is not enough to reduce the cost of capital to *develop* an innovation but also to *commercialize* it, as without commercialization, the enterprise can neither create value nor appropriate it. If the cost of capital for commercialization is high, it works as a disincentive to invest in the development of the innovation. Of course, for an innovating enterprise, earning revenues through sale and licensing of the innovation is a possibility, instead of commercializing it on its own. But, value creation through that process is feasible only when technology markets are mature and functioning well.*

R&D subsidies or tax incentives are ways of reducing cost of capital for developing innovations and India has tried out a variety of instruments for this purpose. As discussed, despite having one of the most liberal tax incentive regimes for R&D, which effectively reduces cost of capital invested for R&D, the Indian corporate sector has not responded in any significant manner by increasing R&D expenditures. One possibility is that, on average, the tax rebates may not fully balance the riskiness of R&D and the delays such a strategy may cause in bringing an innovation to the market. Consequently, firms tend to allocate a larger share of the available resources for technology purchase. Lowering the cost of capital may tilt the balance in favour of R&D but it may also result in more technology purchase as that remains a less risky option. From the policy perspective, it may not always be non-optimal if firms purchase technology and build

---

* As discussed in Chapter 3, some developed economies with mature capital markets are able to provide debt capital for innovation activities and even smaller firms are able to use it, as patents owned by the firm are used as collaterals. The Indian financial markets are yet to evolve to a level where intangibles can be used for raising debt capital.

on the knowledge embedded in it. However, R&D is required to learn better from purchased technology and in that sense, more internal R&D is critical.

There may be another reason for R&D expenditures not showing a significant increase and that relates to the nature of R&D undertaken by Indian enterprises and the reporting of such expenditures. There is ample evidence of the existence of innovation or technology spillovers resulting in firms being able to learn from others' innovation efforts without appropriately paying for it. Some such data has also been shared earlier in the context of India wherein firms report a variety of external sources as critical for knowledge acquisition. It is probable, therefore, that the liberalization processes have enhanced the imitation potential with new technologies coming into the market through imports, FDI, export exposure and patents, and firms are making efforts to benefit from them either by building on them or by inventing around them. However, since these efforts are often not formally organized, they may not have been captured in R&D estimates. This may also be the reason why R&D expenditures have not shown a significant rise over time. One might argue, however, that with a liberal tax regime, firms have the incentive to 'over-report' R&D expenditures and therefore even minor expenses should have been included. But as mentioned in an earlier chapter, the requirements for getting tax benefits are onerous and the transaction costs associated with the bureaucratic processes are high, especially for smaller firms. This may result in under-reporting of R&D expenditures, especially if these are more informal in nature.

Overall, therefore, once we recognize that innovation is not only linked to R&D activities and many other informal and formal activities, including technology purchase and investments in complementary assets, are important for innovation, the issue of cost of capital becomes very important for innovation activities. Given the bureaucratic hassles associated with R&D tax credits, such policies may not be able to compensate for the high cost of capital. Besides, the general cost of capital-related issues discussed above may also be resulting in a lower uptake of tax incentives and subsidy benefits by Indian enterprises. A reduction in cost of capital is critical for all investments, but especially for those related to innovation as they are riskier. In other words, cost of capital-related issues need to be addressed through the entire supply chain—invention, innovation and diffusion. If the relevance of the linkages among the three stages is not recognized, policy instruments to reduce cost of capital only at one stage may not be efficacious, resulting in a less than expected increase in R&D and other innovation activities.

6.2.3. ICT and Other Infrastructure

The emerging role of ICT infrastructure in the innovation process cannot be overemphasized. As discussed, ICT infrastructure has a direct impact on practice innovations as the development as well as adoption of digital innovations depends on access and costs of such infrastructure. Digital technologies are also very critical for product and process innovations. With the convergence of technologies, the role of ICT has become even more important and is being referred to as *innovation technology*. Therefore, a policy focus on availability and inexpensive access to ICT infrastructure is extremely desirable. Insofar as availability of power infrastructure is found to be important for the adoption of ICT and building relevant infrastructure, an enhanced focus on improving availability of power is also quite important.

The telecom infrastructure has grown quite rapidly in India and has facilitated adoption of digital technologies. While that process needs to continue, rapid growth of power infrastructure is a prerequisite for faster adoption of ICT. Else, enterprises have to invest in captive power plants which can be quite expensive for firms, especially the smaller ones. It also needs to be emphasized that high cost of capital also adversely affects investments in ICT and other infrastructure by enterprises.

Good availability of ICT and other infrastructure may also affect to a significant extent, the business conditions that prevail in a region. Moreover, good business conditions were more important for smaller-sized and less productive firms to undertake R&D in response to reforms than for larger-sized and more productive firms. The finding about the higher importance of good business conditions for the smaller and less productive firms is particularly insightful. The importance of good ICT and other infrastructure is also likely to be higher for innovation decisions of smaller firms and firms that are not close to the technology frontier. Typically, such firms have limited resources and/or a higher technology gap to fill and availability of such infrastructure reduces their costs of catching up.

Finally, good ICT and other infrastructure may add to India's attractiveness as a host for FDI and more knowledge-intensive MNC activity. With globalization of research, technology development and production, ICT plays a critical role in a country's effective participation in such networks which are mainly driven by MNCs. With better ICT infrastructure, not only would the ability of Indian firms to participate in these networks get enhanced, the extent and quality of external knowledge flows may also improve.

6.2.4. Higher Education

The discussion in the earlier chapters suggests that low quality of higher education has resulted in HEIs being a very weak link of India's NIS and probably the biggest constraint for innovation-driven long-term growth of the economy. The absence of a large number of good-quality HEIs not only adversely affects our ability to build innovation capabilities, it also has a variety of unintended consequences. According to a recent report, alumni of Indian origin have contributed more than \$1.2 billion to their alma maters in the US since 2000.* In fact, large Indian corporates have also been contributing significantly more to foreign HEIs than they do to domestic ones, probably because such contributions provide them with better branding. For a long time, the better-endowed entities in India—business persons, bureaucrats and politicians—have been sending their children abroad for higher education. While we can feel good about 'brain circulation' and 'brain gain,' brain drain is a reality and with it comes missing out on contributions from alumni as well as corporates!

A significant part of the book focused on the role of higher education in innovation and on some of the key challenges in this domain in the context of India. It was argued that while the labour market linkages between higher education and innovation are the most dominant in India today, even for these linkages there are significant gaps that need to be addressed. In fact, the need to strengthen labour market linkages is more urgent as the higher order links of knowledge generation and enterprise creation would only evolve gradually. Structural

---

* https://timesofindia.indiatimes.com/nri/us-canada-news/pios-have-donated-more-than-1-2-billion-to-us-alma-maters-since-2000/articleshow/65955043.cms

changes in higher education are required in order to make HEIs capable of responding to emerging innovation challenges and effectively play the role they are expected to play in the NIS. Most HEIs are not research-intensive and even where some research gets done, governance-related and other problems undermine their efficacy in getting converted into innovation outcomes. Enhancing research intensity of HEIs is critical for building innovation capability and absorptive capacity, which, the book strongly argues, is essential for any innovation-related policy to be successful. To face competition and exploit contagion conditions, firms in India need such capabilities, but the time horizons of business enterprises are typically short and positive externalities constrain their training efforts. Consequently, HEIs are critical to provide relevant capabilities on a large scale, keeping in mind the long-term needs of the economy.

Currently, HEIs are neither nor do they compete with each other for resources. Therefore, an exclusive focus on policies that try to leverage incentives to generate IP in HEIs does not make much sense. Besides, such a policy takes the focus away from more critical constraints like autonomy of educational institutions and the need to build multiple linkages between the university and industry. Restructuring of HEIs and facilitation of industry-university linkages remain critical. Moreover, other roles that educational institutions can play in the NIS and in building the entrepreneurial ecosystem need to be explicitly recognized. But the HEIs may not be able to perform these roles without any policy impetus to provide autonomy and build an innovation ecosystem. We shall discuss some such possibilities in the next section.

A critical structural gap that ails higher education and makes it less research-intensive is the separation of research and teaching activities that have resulted from a variety of policy initiatives in the past. Linking national labs to universities to create new knowledge systems which are responsive to industry needs has been suggested as a mechanism to make universities more research-focused (Forbes 2017). If done well, this can also correct the anomalies created due to the separation of research and teaching. While this is an interesting idea, it may not bring in multi-disciplinarity to the ecosystem of HEIs, which is also critical for innovation. The New Education Policy announced in 2020 (NEP 2020)* explicitly brings the importance of multidisciplinary education into focus. It also highlights the criticality of improving the quality of higher education and making it more research-intensive. The associated need to change governance structures to enhance faculty and institutional autonomy is also stressed upon. Surprisingly, NEP 2020 also suggests a separation of teaching-intensive and research-intensive universities apart from creating degree-granting colleges. The logic and advantage of such a categorization which again brings in some kind of separation between teaching and research is unclear. To be fair, NEP envisages that a teaching university will also undertake research and a research university some teaching, but why create such categories upfront? Different universities would evolve into less or more research-intensive universities, given resources and other factors at play, as has happened in other parts of the world. But creating a policy-based hierarchy does not seem appropriate; the approach should be to encourage research and combine research and teaching to the extent possible in all HEIs.

---

* For NEP 2020, see: https://www.mhrd.gov.in/sites/upload_files/mhrd/files/nep/NEP_Final_English.pdf

While the ideas envisaged in NEP 2020 have been generally welcomed, it falls short on providing clear guidelines for implementation and leaves several things undefined. For example, Jayal (2020) argues that despite being mentioned seventy times in sixty-six pages, often in conjunction with the term 'holistic', NEP 2020 fails to clearly define 'multi-disciplinary'. The policy document gives an impression that the term is coterminous with the liberal arts curriculum practised in ancient India. In the western context, the term liberal arts encompasses humanities, the social sciences, mathematics and the biological and physical sciences. This needed explicit recognition but NEP 2020 falls short on this count.*

Moreover, it is not entirely clear how and when NEP 2020 will see some action on the ground. As such, the policy only provides a broad direction and any implementation would require collaboration between Centre and the states—education being a concurrent subject wherein both the Centre and the states can make laws. Since many policy suggestions can be seen as encroaching on the decision-making power of the states, an early and smooth implementation seems unlikely. The target of spending 6 per cent of the GDP on education is laudable but it is unclear if the new policy implementation will not be constrained by availability of funds as was the case with some earlier policies.

Even if India is able to get some of these things right in a few institutions, the impact is unlikely to be high as the larger higher education system will still suffer from the old rules of governance which constrain the building of an innovation culture. The state in its wisdom has created many new IIMs and IITs that struggle on various fronts, while the older IITs and IIMs, better in terms of their academic culture, remain

---

* Jayal (2020) provides an excellent discussion on this issue.

sub-scale. As discussed in an earlier chapter, these institutions still do not have autonomy on several fronts including deciding their salary structure. NEP 2020 does not provide any clarity on the nature of autonomy the HEIs would enjoy after the creation of the Higher Education Commission of India (HECI) that will replace the UGC. The only thing that NEP 2020 mentions is that post the creation of HECI, both public and private institutions would be subject to the same accreditation and academic standards. How these institutions would be governed, what standards would be applied and how is still unknown. Besides, there is a fair bit of ambiguity about how various regulatory institutions would be formed and their composition (Jayal 2020). Irrespective of what policy options the nation adopts, it may not be feasible to quickly create large multidisciplinary HEIs.

NEP 2020 also envisages opening up the higher education sector to foreign universities, allowing them to set up campuses in India, provided they are among the top 100 in the world. This is expected to improve the quality of higher education in India. Currently, participation of foreign universities in India is limited to collaborative twinning programmes wherein they share faculty with partnering institutions and offer distance education. Many foreign education providers have such arrangements in India. It is not clear if the new policy would result in the best universities setting up campuses in India as the top universities did not seem very enthusiastic about making an entry into India when the earlier government was contemplating a similar initiative. The new law based on NEP 2020 that would include details of how foreign universities will operate in India, would partly decide the willingness of top global universities to set up campuses in India. Therefore, even assuming the willingness of top foreign universities to set up campuses in India, foreign entry-based improvement in higher education is also unlikely to happen in the near future.

One possible policy option is to develop a network of diverse HEIs that are co-located in a city cluster. Policy initiatives that can facilitate city-specific collaborations can potentially create opportunities for multidisciplinary learning and innovation. This collaboration can take the form of research projects, project courses, incubation and/or acceleration of start-ups across institutions, and so on. Similarly, there are possibilities of offering courses that combine science, technology, law, design, management, humanities and social science inputs. Some efforts in this direction might facilitate the co-evolution of networking institutions as envisaged in some versions of the NIS literature. The structural flaws of stand-alone single-discipline institutions can potentially be corrected reasonably quickly, at least partly, by such city cluster-based networks and collaborations. Interestingly, NEP 2020 refers to HEI clusters or knowledge hubs that can potentially create 'vibrant multi-disciplinary environments' but does not provide any specifics on the potential policy options or direction. Several city clusters in India have good HEIs in specific disciplines but do not collaborate, presumably because of lack of incentives and because institutional boundaries are difficult to transcend.* An active network of these institutions can simulate the ecosystem of a large multi disciplinary university. Institutional silos are difficult to break but a policy instrument on the lines of the SETsquared partnership† initiative in the UK to bring

---

* Apart from the metros, cities like Ahmedabad, Bengaluru, Pune and Hyderabad would certainly qualify to be part of such clusters. In Ahmedabad, an effort is being made to bring together the local HEIs to explore collaboration possibilities

† For details, see https://www.setsquared.co.uk/

together five universities in a region may be worth a try. As these collaborative efforts evolve, local industry is bound to get involved with the local HEIs for mutually beneficial partnerships. One possibility of initiating such partnerships could be through mission-oriented programmes which can not only bring together HEIs in the city cluster but also industry and other stakeholders. We shall revert to this issue in the next section. Irrespective of what mechanism we use to reinvigorate the HEIs in India to build innovation capabilities, public funding of research and teaching in HEIs and their autonomy will remain critical for building such capabilities.*

6.2.5. Skill Gaps

Just like the quality of higher education leaves much to be desired, the skill gaps in India remain very high. The need of well-trained persons to implement innovations in product, process and practice spaces is obvious. A variety of policy initiatives to build skills have been tried over the years without much success. As discussed, the most recent Skill India programme has also not been an exception. If the skilled workers are not readily available, in-house training has to be undertaken which can be expensive. Besides, the fear of losing trained workers to competitors and high cost of capital can deter business enterprises from investing in training. Still, as the data discussed in an earlier chapter shows, a large proportion of enterprises do undertake such training but these investments may not be adequate. The training expenses as a proportion of

---

* It has been suggested that CSR funds can potentially complement public funding, especially for mission-driven R&D if the laws are appropriately modified (Economic Survey, 2017–18).

sales have also not risen consistently in recent years. The market failure resulting in underinvestment in training by industry was sought to be corrected through PPPs in skill development. NSDC is one such organization, set up as a not-for-profit company and as a PPP. The current model does not seem to be working and can benefit from other PPP models. An arms-length relationship between NSDF and NSDC is desirable to avoid any conflict of interest and at the same time for NSDC, autonomy and accountability should go together as is the case in Infuse Ventures*. A more direct and active participation of industry in identifying skill gaps and appropriate programmes to fill them is critical. The discussion in Chapter 4 suggests that presumably, the current bureaucratic structure and other organizational constraints are keeping the private sector away from playing a more active role, as financiers and also in identifying and addressing skill needs. It is critical to think about an appropriate organizational model to create and sustain a PPP in the vocational training space. Once an incentive-compatible organizational structure is in place, the private sector may also contribute to help set up a large number of VET institutions. Many of these can be run by the private sector to provide long-term competency-based courses according to the emerging requirements of industry. Another possibility is that the NSDF-NSDC part fund on-site training programmes within the companies, subject to some quality outcome standards.

---

* To enhance accountability, Government of India (2016) recommends that the NSDC should operate as a Non-Banking Financial Company (NBFC), as it is already financing different stakeholders involved in skill development schemes. It will then be subjected to audit on outcome basis.

6.3. Innovation Policy: Some Specific Focus Areas

In the last section, a few generic issues relating to innovation policy were discussed. These issues are in line with the broad conceptualization of innovation policy that includes all policy instruments which can influence the decision of an economic entity with respect to innovation efforts and outcomes. Many of the policies discussed were macro policies that impinge on innovation efforts and outcomes. In this section, we focus on a few specific policy areas, especially those which have been under discussion in recent years. This is done in the context of the arguments articulated in the earlier chapters.

6.3.1. Start-ups and Incubation

Innovation and entrepreneurship are buzzwords in India today. Analyses of their potential roles not only permeate the discussion around *Start-up India* and *Standup India* but also figure in discussions on *Make in India* and *Skill India* initiatives as the latter two are expected to benefit from innovation-driven entrepreneurship, especially new ventures. It is imperative that the synergies among these major initiatives of the federal government are systematically reaped. While a detailed understanding of various linkages across these programmes would require more rigorous analysis, in line with our earlier discussion, it would be useful to focus on the role of the state and HEIs in creating an ecosystem where technology entrepreneurship can flourish.

Externalities play an important role in creating and nurturing an innovation and entrepreneurial ecosystem. The ecosystem comprises individual inventors, large corporates, HEIs, entrepreneurs, VCs, angel investors, consumers, vendors, employees, service providers and governments.

Start-ups and VCs not only benefit from the existence of their peers but also by the actions of others through what is referred to as a spillover effect. The role of HEIs as a source of research that can be taken to the market, skills as well as entrepreneurs, has been stressed in the context of vibrant entrepreneurial, ecosystems in the West. One aspect that has remained under-emphasized is the nature of the universities in these innovative clusters. These are large multidisciplinary research-intensive entities with 'spaces' for different disciplines to interact. Innovation and entrepreneurship are multidisciplinary activities; the interaction of science, engineering, design, management, humanities and other disciplines provides a fertile ground for nurturing and growing innovative ideas. As noted, unlike the West, India does not have universities where such interactions happen. In a few universities where such possibilities exist, the silo culture and lack of interaction spaces come in the way of undertaking any interesting cross-disciplinary experiments.

HEIs host almost all the state-supported incubators in India. Most of the incubation centres are located in technology institutions where the start-ups are not able to get good business-related support; the same is true of start-ups in business schools with respect to technology support. Moreover, start-ups in technology and management schools find it difficult to access design and other inputs that are relevant to understand the market and the user. In the short run, it does not seem possible to create such research-intensive and multidisciplinary universities in India. But as mentioned, there are several city clusters where high-quality HEIs representing different disciplines coexist. In many of these clusters other elements of the entrepreneurial ecosystem are also present. Efforts to bring these institutions together around innovation and entrepreneurship-related activities—joint courses, research, co-incubation, etc.—can be a potent force to unleash innovation-driven entrepreneurship.

As discussed earlier, one method could be to tie part of the innovation and incubation-related funding to networking between institutions as is done in the SETSquared partnership in the UK. Typically, government agencies support incubation centres in specific HEIs which specialize in different domains—design, management, technology, etc. The policy of supporting incubation centres in specific institutions is administratively easier but does not exploit spillovers and economies of scale and scope. Policy instruments that provide incentives for joint activities among these institutions can go a long way in creating the requisite entrepreneurial culture. It is desirable to support incubation centres for a cluster of institutions rather than for an individual institution. Appropriate organizational structures cutting across boundaries of participating institutions will have to be worked out but if done right, the positives are many. Apart from bringing together multiple disciplines and thereby improving innovation and incubation outcomes, such an effort will also help in reaping economies of scale and scope in incubation activities. Most incubation centres are sub-scale and therefore, financially unsustainable. This also results in incubators not being able to hire and retain a good management team, adversely affecting the quality of incubation. With better networking across partner HEIs from various disciplines and other elements of the ecosystem—user industries, angels, VCs, mentors—in the city cluster, the vibrancy of the innovation and entrepreneurial ecosystem would also improve.

6.3.2. PPPs for Financing Start-ups

We have argued in different parts of the book that innovation requires sustained financial support and a short-term perspective of financial markets is not conducive to the growth of innovation-driven start-ups. Private VC firms typically do not provide patient capital needed for radical innovations. In any case, the focus of VCs on early and profitable 'exit' results in limited investments by them in early-stage technology-driven start-ups. Such a tendency is more prominent in economies like India where financial markets are not mature enough to create several opportunities of exit. Apart from research grants, governments worldwide are also providing entrepreneurial capital. As mentioned in Chapter 4, the Indian government (Department for Promotion of Industry and Internal Trade) has also announced an INR 10,000 crore FFS for start-ups that is being implemented by SIDBI. The data reported earlier shows that the disbursements from this FFS have been rather low and a variety of problems have constrained the scheme's implementation.

The case of Infuse Ventures discussed in Chapter 5 provides some insights that might be useful for the SIDBI-anchored initiative. In fact, the insights are useful for all sectors where technology and market risks are high. As discussed, Infuse is a PPAP wherein government funds were leveraged to raise resources from the private sector and invested in clean energy start-ups through an independent investment committee hosted by an academic institution. This can potentially be a model for deploying a new start-up fund for other sectors. Available government resources can be used as a seed to create venture funds for sectors where the market and technological risks apart from other types of market failures are high. Such a model to provide entrepreneurial capital for early-stage entities can bring in advantages of 'syndication' through its PPP element and also reduce chances of government failures.

The details of the structure used for Infuse Ventures have already been discussed and there is no need to repeat the same here. Essentially, it created incentive compatibility for all the key stakeholders in the clean energy sector. Each segment of the private players benefited as the risks, uncertainties and transaction costs they typically face were reduced through this intervention apart from reducing the risks associated with a weak ecosystem. A similar role can be played by FFS as anchor but subordinated investor. Besides, standardization of documents, provision of draw-down and co-investment facility, an arms-length implementation to maintain autonomy of the investment adviser, enablement of the use of CSR funds, availability of venture debt and acceleration funds are insights that are very useful for FFS as they have proved to be effective in the context of Infuse and other similar contexts.

Moreover, this kind of funding initiative also has significant benefits for the academic incubator managing the fund, if it is able to build appropriate capabilities required to anchor such an intervention. Thus, such arrangements can help the academic incubators build scale and enhance their financial viability, which is an important concern with the incubator support provided by the government. Infuse also gives an interesting example of providing patient capital by combining developmental, strategic and commercial capital and can be used as a model for similar PPP arrangements for sectors where market failures are high but high-tech innovation-driven start-ups can make significant social impact. Broadly, such arrangements are likely to be useful for all sectors where regulatory, market and technological uncertainties are high with private capital shying away and the state not being a good substitute as it is not an experienced technology investor. Once a few such arrangements become successful like Infuse, a process may get initiated wherein institutional investors like pension funds, sovereign wealth funds and family offices may contribute to the creation of such funds as they are not keen to realize returns very quickly. The Infuse experience also suggests that having stakeholders that complement each other's capabilities and contributions facilitates success for such endeavours.

The Infuse experience also suggests that if resources are available, the ideal design for a PPP fund like Infuse for a technology-intensive sector with a wide social impact would be a long-dated multi-stage dual-mission structure. Since the success of interventions in early-stage high-tech social impact finance is likely to be dependent on availability of growth equity, one cannot take care of the early-stage financing gaps without addressing problems of late-stage financing. So, PPP funds that wish to address a high-tech space with social impact should ideally have a dual mission of early-stage and growth capital intervention. This would imply a larger fund size as the mandate would be to invest in both stages—early as well as growth. Such a mandate would also necessitate a long-dated fund, with a duration of fifteen–twenty years (instead of the usual eight–ten years), allowing the provision of a line-of-equity to its portfolio ventures (linked to performance) that extends support from early to growth stages. The FFS has the potential to create such a fund.

More recently, an FFS has also been proposed for MSMEs under the COVID-19 Stimulus Package. As is the case of the earlier FFS, the Infuse experience has important lessons. One, the government needs to act as subordinated capital if it wants market players to actively contribute. Else crowding-in of capital may not happen easily, especially in COVID-like situations. Two, it is imperative that the government facilitates faster deployment as the fundraising cycle can be very long for fund managers to raise matching amounts.

In the absence of faster deployment, the whole purpose may get defeated. This, as discussed, was the case with SIDBI-DIPP funds where deployment got delayed due to fund-raising delays. Faster deployment can be done if, while the funds are being raised, the government considers working with existing incubators and funds to directly deploy without any additional commitments from other investors. In effect, the government needs to decide if the primary objective of the MSME FFS is to catalyse MSMEs or generate returns. These are two divergent objectives as was discussed in the context of Infuse and, at least in a COVID-like situation, rejuvenating MSMEs and even start-ups should probably be the primary objective.

6.3.3. PPP in Research Funding: Mission-Driven Approach

Most governments across the world support research and finance innovation. Many of these initiatives have been quite successful. Studies reviewed earlier show that the efficacy of state initiatives not only depends on the extant ecosystem but the nature of intervention. Studies suggest that financing of innovation is likely to be more successful in situations where: (i) the public sector spends large sums on education and research in those emerging sectors where high capital intensity and technological or market risks result in underinvestment by the private sector; (ii) large firms reinvest their profits in human capital and R&D; (iii) the tax policy rewards long-run investments rather than short-run capital gains; and (iv) contestable markets along with a rigorous competition policy does not allow successful incumbents to become lazy. But what should be the nature of state financing? We have already discussed the potential role of the government in financing innovation-driven start-ups through PPP funds and the issues that might be relevant to consider while providing tax benefits for R&D. Here we explore some policy options in direct research funding.

Several studies, many of them referred to earlier, have pointed out that technology intensity of manufacturing and other activities in India is rather low. Therefore, India needs a stronger presence in sectors that are technologically dynamic worldwide. Given the issues discussed earlier around the need for contestable markets, trade policy-based protection of such industries is not likely to be the appropriate option. Direct mission-driven support would probably make more sense. On the face of it, the government seems to be exploring this option actively. The introduction of SBIR-like programmes, discussed in Chapter 4, and the programme called IMPRINT (**IMP**acting **R**esearch **IN**novation and **T**echnology) launched in 2015 are examples of such initiatives. The IMPRINT scheme is a multi-stakeholder mission programme with a focus on supporting translational research by HEIs engaged in science and technology-related areas.* The programme identified ten domains: advanced materials, energy, environment and climate change, healthcare, information and communication technology, manufacturing, nanotechnology hardware, security and defence, and water and river systems. Challenges are identified for each domain, seeking proposals from HEIs. Apart from the Ministry of Human Resource Development (MHRD), twenty-seven user ministries and departments are involved in the process of identifying challenges and specific HEIs anchor different challenges. However, active participation of industry in defining challenges and implementing solutions

---

* For details of the programme, visit: https://imprint-india.org/ and https://imprint-india.org/imprint-2

is not envisaged, at least not articulated in any clear manner. It is possible that some panels reviewing proposals may have industry representation but it does not seem to be an explicit part of the programme design. There is some scope for their participation in piloting the projects but it is apparently not utilized much. The entire exercise seems to be driven by the government and select HEIs which is surprising as the role of industry is critical in translational research.

An internal review of the programme apparently recognized this gap and version two of the programme (IMPRINT-2) was to be more inclusive and demand driven. But the modalities of seeking proposals and commercialization still do not require the participation of industry. Building incubation linkages for these projects is, however, mentioned but it is not clear how this has been implemented. While no review of IMPRINT is available, prima facie, the programme seems to have the same problems that were faced by the SBIRI and BIPP initiatives discussed in Chapter 4. Besides, the funding through such programmes also does not ensure government procurement.

Interestingly, the *Economic Survey* of 2019–20* also argues that India should focus on manufacturing 'network products' that have globally fragmented production processes, typically anchored and controlled by MNCs. Participation in the global production networks of such products that include computers, electronic and electrical equipment, telecommunication equipment, automotive and so on is expected to create jobs and export performance as was evidenced in China. In doing so, the *Economic Survey* posits the idea of integrating 'Assemble in India for the world' into Make in India! Many of these network products are high-tech products with the scope of facilitating the building of

---

* https://www.indiabudget.gov.in/economicsurvey/

dynamic technological capabilities in the nation. However, the potential of assembling and later making high-tech products in India would require the presence of all the enabling policies referred to earlier in the chapter including contestability, low cost of capital, better access and lower cost of ICT, reforming higher education and so on. Some of the missions included in the SPRINT and SBIRI-like programmes will also need to be dovetailed with the Make in India effort. In fact, it is critical that such linkages are explicitly explored and acted upon with alacrity. A Make in India-like programme can only succeed in building innovation capabilities if all the three elements of the NIS—government, industry and academia—are actively involved.

Any mission-driven programme for building innovation capacity in India needs to take cognizance of insights from the analysis in earlier parts of the book and from two recent reviews: one of the SBIR programme in the US (Wessner 2008) and the other of mission-oriented innovation policies (Mazzucato 2018).

Applications for SBIR funding are made in response to a call for proposals that describe the types of projects that will be funded. The topics and sub-topics for funding are identified through three kinds of models (Wessner 2008):

1. *Acquisition-oriented topics* are linked to the acquisition priorities for the awarding agency (e.g., NASA and Department of Defence). The objective is to target proposed R&D towards projects that will result in technologies that can eventually be acquired by the agency for its use.

2. *Management-oriented topics* serve the research agendas of different agencies like the various divisions of the National Science Foundation (NSF) and the Department

of Energy (DoE). These agencies do not procure technology but such an approach enhances buy-in of agency staff for SBIR.

3. *Guideline-oriented topics* only indicate broad areas of interest and do not act as delimiters (which is the case with the first two) for what projects can be applied for. The National Institutes of Health (NIH) uses this approach and explicitly mentions that it primarily supports investigator-driven research. Consequently, applications on any topic are potentially acceptable.

The approach followed by the IMPRINT and SBIRI programmes in India seem closer to the management-oriented approach. The other two approaches are not used. Wessner (2008) argues that all the approaches have their advantages and disadvantages. For example, an investigator-driven approach may attract better science projects but this may not be suitable for an agency's acquisition needs. But narrowly defined needs can help focus on a specific set of technologies. The SBIR programme is found to be effective partly because a 'one-size-fits-all' approach has not been imposed. The review concludes that the acquisition-oriented approach seems appropriate for agencies where technology acquisition is the primary objective. It is argued that the management-oriented model used by NSF is inappropriate as it does not necessarily serve the interest of science. Available evidence suggests that the investigator-driven model seems to be working effectively for NIH. Given diverse needs, a hybrid model is suggested which is also being used in nations like Sweden wherein half of research funding is allocated to agency programmes focused on the research needs of specific industries and the other half is allocated to proposals for research initiated by companies.* Furthermore, the review recommends selection of topics for SBIR in a more transparent manner and the use of a bottom-up rather than a top-down process, with more industry and investigator participation.

Given the limited resources for such endeavours in India, unlike in the US where SBIR programmes get significant and sustained funding, a hybrid model would also be more appropriate. Irrespective of the approach used, industry participation in both kinds of programmes is critical given the absence of skilled personnel with first-hand industry experience in government agencies in India. The presence of very skilled programme managers is one of the critical reasons for the success of the SBIR initiative. These insights are likely to be useful for the National Research Foundation (NRF) envisaged in NEP 2020.

The discussion so far focused on PPPs or quasi-PPPs engendered by SBIR-like endeavours that have been implemented with a different flavour in India through IMPRINT, SBIRI, BIPP and other initiatives. These can be seen as missions in a somewhat narrow sense wherein specific areas get defined for research efforts. At a broader level, mission-oriented initiatives can be seen as a set of systemic policies that seek to draw on state-of-the-art knowledge to solve large societal problems. Mission-oriented policies therefore not only try to correct market failures but proactively shape and create markets (Mazzucato 2018). The possibility of government failure is therefore high in such initiatives. One way to avoid such government failures is to involve and build dynamic links between different entities of the innovation ecosystem—enterprises, HEIs, research organizations, institutions engaged

---

* For details, see Wessner (2008) and visit https://www.vinnova.se/en/ to know more about the Vinnova initiative.

in innovation financing (both public and private) and other institutions including civil society and political organizations depending on the nature of the mission. Mazzucato's (2018) review suggests that for success, missions should be defined in a pragmatic manner and should create a long-term public agenda with active participation of various stakeholders. The mission should be feasible, address a societal demand and draw on existing private and public resources including those relating to science and technology. In addition, these should be enabled with complementary policies and receive broad and continuous political support. Admittedly, it is quite onerous to achieve these conditions, especially in functioning political democracies. But it is useful to keep them in mind if one wishes to adopt a mission-driven approach. The least one can try to do is to combine demand pull and technology push approaches by involving all key constituents of the NIS. In the current endeavours, such participation seems to be missing, both in setting the agenda as well as in implementing it.

6.4. Concluding Observations

It is generally argued that the critical contribution of the government is to provide stable policies that make doing business easier. Such policies are expected to spur innovation. While there has been a lot of scepticism about the government being a direct source of innovation, many recent empirical studies have painstakingly documented the critical role that the state has played in fostering innovation and entrepreneurship, even in places like Silicon Valley. It is being recognized that both state-supported research as well as entrepreneurial capital are necessary (though not sufficient) conditions for innovation. This chapter has discussed in detail the conditions under which state intervention is likely to be more successful.

The innovation needs of India are diverse and complex. A wide variety of policies have been announced in recent times to make the country an innovative self-reliant economy without clearly stating the underlying premises or defining the broad strategic policy framework. Building technological capabilities is critical for the long-term economic growth of a nation. It is critical that innovation policies are seen in a broader perspective, recognizing the linkages across policies that influence innovation activities and outcomes. Consistency across such policies needs to be ensured and all constituents of the NIS should actively participate in the endeavour to build innovation capabilities. Start-ups supported by VCs along with the support of incubators and the rest of the ecosystem are increasingly seen as critical for fostering innovative enterprises. They are also expected to solve the problems of employment and growth. But many elements of the policy framework are still not in place for the innovative start-up-led process to make a significant dent. Besides, historical evidence suggests that in a healthy innovation ecosystem, while VCs and start-ups are important contributors for stimulating innovation, they can never be a substitute for other sources of innovation, namely vibrant higher education institutions and R&D undertaken in the large corporate sector—both public and private. Various policy options available to India to achieve some of these objectives have been discussed in the book. It is hoped that some of these will be given serious thought and, if found relevant, implemented. But we still need to work towards building a scientific temperament in the country for a sustained innovation-driven economy. Antediluvian thought processes are not conducive to science and technology-based economic growth. Building of modern autonomous and liberal educational institutions and policy options discussed here can partly contribute to the building of such a scientific temper. But these are unlikely to be sufficient to make the relevant societal changes that are critical to make scientific temper more pervasive. That is a problem much more complex than what the book has tried to explore. Such an exploration is urgently needed but surely requires better minds.

References

Abernathy, W.J. and Utterback, J.M. (1978). Patterns of Industrial Innovation, *Technology Review*, 80(7), pp. 41–47.

Aggarwal, A. and Chawla, S. (2014). *Promoting Innovation through Public Private Partnership: An Assessment of the SBIRI/BIPP Programmes*, Wadhwani Foundation.

Aghion, P., Bechtold, S., Cassar, L. and Herz, H. (2014). 'The Causal Effects of Competition on Innovation: Experimental Evidence', *NBER Working Paper No. 19987*, National Bureau of Economic Research, Cambridge.

Aghion, P., Bloom, N., Blundell, R., Griffith, R. and Howitt, P. (2005). 'Competition and Innovation: An Inverted-U Relationship', *The Quarterly Journal of Economics*, 120(2), pp. 701–28.

Aghion, P., Blundell, R., Griffith, R., Howitt, P. and Prantl, S. (2006). 'The Effects of Entry on Incumbent Innovation and Productivity', *NBER Working Paper No. 12027*, National Bureau of Economic Research, Cambridge.

Aghion, P., Blundell, R., Griffith, R., Howitt, P. and Prantl, S. (2009). 'The Effects of Entry on Incumbent Innovation and Productivity', *Review of Economics and Statistics*, 91(1), pp. 20–32.

Akkoyunlu, S. (2013). 'The Correlation between the Level of Patent Protection and International Trade', *Working Paper No. 2013/36*, The National Centre of Competence in Research, Swiss National Science Foundation.

Allen, R.H. and Sriram, R.D. (2000). 'The Role of Standards in Innovation', *Technological Forecasting and Social Change*, 64(2-3), pp. 171–81.

Amiti, M. and Khandelwal, A.K. (2013). 'Import Competition and Quality Upgrading', *Review of Economics and Statistics*, 95(2), pp. 476–90.

Amsden, A.H. (1989). *Asia's Next Giant: South Korea and Late Industrialization*, New York: Oxford University Press.

Anand, A. (2020). 'Economic Policy Reforms, Foreign Direct Investment and the Patterns of MNC Presence in India: Overall and Sectoral Shares', *Working Paper No. 493*, Centre for Development Studies.

Arora, A. and Gombardella, A. (1990). 'Complementarities and External Linkages: The Strategies of Large Firms in Biotechnology', *Journal of Industrial Economics*, 38(4), pp. 361–79.

Arora, A. (1991). 'Transferring Tacit Knowledge in Technology Transfer: How Can Intellectual Property Rights Legislation Help the Industrialising Countries?', *Mimeo*, Stanford University.

Arrow, K.J. (1962). 'Economic Welfare and the Allocation of Resources for Invention', In R.R. Nelson (Eds.), *The Rate and Direction of Inventive Activity*, Princeton: Princeton University Press.

Athreye, S. and Prevezer, M. (2008). 'R&D Offshoring and the Domestic Science Base in India and China', *Working*

Paper No. 26, Centre for Globalisation Research School of Business and Management.

Athreye, S. and Puranam, P. (2007). *India and China: Their Role in the Global R&D Economy*, in MEIDE conference at Renmin University, Beijing.

Baldwin, J.R. and Gu, W. (2004). 'Trade Liberalization: Export-Market Participation, Productivity Growth, and Innovation', *Oxford Review of Economic Policy*, 20(3), pp. 372–92.

Banerjee, R. and Muley, V. (2008). Report on Engineering Education in India, Observer Research Foundation, *Mimeo*, Delhi.

Banerji, A. and Suri, F.K. (2017). 'Patents, R&D Expenditure, Regulatory Filings and Exports in Indian Pharmaceutical Industry', *Journal of Intellectual Property Rights*, 22, pp. 136–45.

Baregheh, A., Rowley, J. and Sambrook, S. (2009). 'Towards a Multidisciplinary Definition of Innovation', *Management Decision*, 47(8), pp. 1323–39.

Bas, M. and Paunov, C. (2018). 'The Unequal Effect of India's Industrial Liberalization on Firms' Decision to Innovate: Do Business Conditions Matter?', *Journal of Industrial Economics*, 66(1), pp. 205–38.

Basant, R. (1990). Regional Overview, In S. Gamser, H. Appleton and N. Carter (Eds.), *Tinker, Tiller, Technical Change*, London: Intermediate Technology Publications.

Basant, R. (1993). 'R&D, Foreign Technology Purchase and Technology Spillovers in Indian Industry: Some Explorations', *Working Paper No. 8,* United Nations University Institute of New Technologies, Maastricht.

Basant, R. (2000). 'Corporate Response to Economic Reforms, *Economic and Political Weekly*, 35(10), pp. 813–22.

Basant, R. (2002). 'Knowledge Flows and Industrial Clusters: An Analytical Review of Literature', *IIMA Working Paper No. 2002-02-01*, Indian Institute of Management Ahmedabad.

Basant, R. (2006). 'ICT Adoption, Skills, Organizational Change and Firm Performance: Some Insights from Indian Cases', Indian Institute of Management, *Mimeo*.

Basant, R. (2007). 'ICT Adoption and Productivity Gains in Indian Manufacturing', *Issue Brief No. 10*, Observer Research Foundation, New Delhi.

Basant, R. (2011). 'Intellectual Property Protection, Regulation and Innovation in Developing Economies: The Case of Indian Pharmaceutical Industry', *Innovation and Development*, 1(1), pp. 115–33.

Basant, R (2018). 'Exploring Linkages between Industrial Innovation and Public Policy: Challenges and Opportunities', *Vikalpa,* 43 (2), pp. 61–76.

Basant, R. (2021). 'Challenges of Growth and Productivity—Innovation Policy in India', In Debroy, B., C. Raja Mohan and Ashley Tellis (Eds.), *Grasping Greatness: Making India a Leading Power*, Penguin (*Forthcoming*).

Basant, R. and Fikkert, B. (1996). 'The Effects of R&D, Foreign Technology Purchase, and Domestic and International Spillovers on Productivity in Indian Firms', *Review of Economics and Statistics*, 78(2), pp. 187–99.

Basant, R. and Majumder, S. (1997). 'Trade Regimes and Productivity: Exploring the Impact of Tariff Policy on Firm Level Technology Strategies', *IIMA Working Paper No. 1997-1418*, Indian Institute of Management Ahmedabad.

Basant, R. and Chandra, P. (2002). 'Building Technological Capabilities in a Liberalizing Developing Economy: Firm

Strategies and Public Policies', *Economics of Innovation and New Technology*, 11(6), pp. 399–421.

Basant, R., Commander, S.J., Harrison, R. and Menezes-Filho, N. (2006). 'ICT Adoption and Productivity in Developing Countries: New Firm Level Evidence from Brazil and India', *IZA Discussion Papers, No. 2294*, Institute for the Study of Labour (IZA), Bonn.

Basant, R. and Chandra, P. (2007a). 'Role of Educational and R&D Institutions in City Clusters: An Exploratory Study of Bangalore and Pune Regions in India', *World Development*, 35(6), pp. 1037–55.

Basant, R. and Chandra, P. (2007b). 'University-Industry Links and Enterprise Creation in India: Some Strategic and Policy Issues', In S. Yusuf and K. Nabeshima (Eds.), *How Universities Promote Economic Growth*. Washington DC: World Bank.

Basant, R. and Mukhopadhyay. P. (2010). 'Arrested Virtual Cycle? Higher Education and High-technology Industries in India', In Yifu Lin J. and Pleskovic, B. (Eds.), *Annual World Bank Conference on Development Economics 2009, Global: People, Politics, and Globalization.* Washington DC: World Bank.

Basant, R. and Mani, S. (2012). 'Foreign R&D Centres in India: An Analysis of their Size, Structure and Implications', *IIMA Working Paper No. 2012-01-06*, Indian Institute of Technology Ahmedabad.

Basant, R. and Cooper, S. (2016). 'Contrasting Models of Incubation for Enterprise Creation: Exploring Lessons for Efficacy and Sustainability from Higher Education Institutions in India and the United Kingdom', *IIMA Working Paper No. 2016-02-05*, Indian Institute of Management Ahmedabad.

Basant, R. and Srinivasan, S. (2016). 'Intellectual Property Protection in India and Implications for Health Innovation: Emerging Perspectives', *Innovation and Entrepreneurship in Health*, 3, pp. 57–68.

Beneito, P., Rochina-Barrachina, M.E. and Sanchis, A. (2014). 'Patents, Competition, and Firms' Innovation Incentives', *Industry and Innovation*, 21(4), pp. 285–309.

Blanco, M.I. and Rodrigues, G. (2009). 'Direct Employment in the Wind Energy Sector: An EU Study', *Energy Policy*, 37(8), pp. 2847–57.

Blind, K. (2013). 'The Impact of Standardization and Standards in Innovation', *Nesta Working Paper 13/15*, Handbook of Innovation Policy Impact, Edward Elgar Publishing.

Bogers, M. and West, J. (2012). 'Managing Distributed Innovation: Strategic Utilization of Open and User Innovation', *Creativity and Innovation Management*, 21(1), pp. 61–75.

Branstetter, L.G., Fisman, R. and Foley, C.F. (2006). 'Do Stronger Intellectual Property Rights Increase International Technology Transfer? Empirical Evidence from US Firm Level Panel Data'. *Quarterly Journal Of Economics*, 121(1), pp. 321–49.

Briggs, K. (2013). 'Does Patent Harmonization Impact the Decision and Volume of High Technology Trade?', *International Review of Economics and Finance*, 25(C), pp. 35–51.

Busom, I. (2000). 'An Empirical Evaluation of the Impact of R&D Subsidies', *Economics of Innovation and New Technology*, 9(2), pp. 111–48.

Bustos, P. (2011). 'Trade Liberalization, Exports, and Technology Upgrading: Evidence on the Impact of

MERCOSUR on Argentinian Firms', *American Economic Review*, 101(1), pp. 304–40.

Cardosa, A. and Duarte, A.P. (2017). 'The Impact of the Chinese Exchange Policy on Foreign Trade with European Union', *Brazilian Journal of Political Economy*, 37(4), pp. 870–93.

Cedrick, B.Z.E. and Long, P.W. (2017). 'Investment Motivation in Renewable Energy: A PPP Approach', *Energy Procedia*, 115, pp. 229–38.

Centre for Technology, Innovation and Economic Research (2019). *Technology and Innovation in India*, CTIER Handbook.

Chakrabarti, A. and Bhaumik, P.K. (2009). 'Internationalization of Technology Development in India', *Journal of Indian Business Research*, 1(1), pp. 26–38.

Chaminade, C. (2010). 'Are Knowledge Bases Enough? A Comparative Study of the Geography of Knowledge Sources in China (Great Beijing) and India (Pune)', *Working Paper No. 2010/13, CIRCLE*, Lund University.

Chandra, P. (1995). 'Technology Characterization: Explaining a Few Things', *Mimeo*, Indian Institute of Management Ahmedabad.

Chandra, P. (2017). 'Investment Analysis and Portfolio Management', McGraw-Hill Education.

Cohen, W. and Levin, R.C. (1989). 'Empirical Studies on Innovation and Market Structure', in R. Schmalensee and R.D. Willig (Eds.), *Handbook of Industrial Organisation*, Vol. II, Elsevier Science Publishers.

Cohen, W.M. and Levinthal, D.A. (1989). 'Innovation and Learning: The Two Faces of R&D', *Economic Journal*, 99(397), pp. 569–96.

Cohen, W.M. and Levinthal, D.A. (1990). 'Absorptive Capacity: A New Perspective on Learning and Innovation', *Administrative Science Quarterly*, 35(1), pp. 128–52.

Correa, J.A. (2012). 'Innovation and Competition: An Unstable Relationship', *Journal of Applied Econometrics*, 27(1), pp. 160–66.

Crespo, N. and Fontoura, M.P. (2007). 'Determinants of FDI Spillovers—What Do We Really Know?', *World Development*, 35(3), pp. 410–25.

Crossan, M.M. and Apaydin, M. (2010). 'A Multi-Dimensional Framework of Organizational Innovation: A Systematic Review of the Literature', *Journal of Management Studies*, 47(6), pp. 1154–91.

CTIER (2018), *CTIER Handbook: Technology and Innovation in India 2019*, Centre for Technology Innovation and Economic Research, Pune.

CTIER (2020), *CTIER Handbook: Technology and Innovation in India 2021*, Centre for Technology Innovation and Economic Research, Pune. (Forthcoming)

Dahlman, C.J., Ross-Larson, B. and Westphal, L.E. (1987). 'Managing Technological Development: Lessons from Newly Industrialising Countries', *World Development*, 15(6), pp. 759–75.

Das, P. (2004). 'Economic Liberalisation and R&D and Innovation Responses of Indian Public and Private Sector Industries', *International Journal of Management and Decision Making*, 5(1), pp. 76–92.

David, D., Gopalan, S. and Ramachandran, S. (2020). 'The Start-up Environment and Funding Activity in India', *ADBI Working Paper No. 1145*, Tokyo: Asian Development Bank Institute.

Dechenaux, E., Goldfarb, B., Shane, S. and Thursby, M. (2008). 'Appropriability and Commercialization: Evidence from MIT Inventions', *Management Science*, 54(5), pp. 893–906.

Del Rio, P. and Burguillo, M. (2009). 'An Empirical Analysis of the Impact of Renewable Energy Deployment on Local Sustainability', *Renewable and Sustainable Energy Reviews*, 13(6-7), pp. 1314–25.

Dodgson, M. and Gann, D. (2010). *Innovation – A Very Short Introduction*, Oxford University Press.

Dosi, G. (1982). 'Technological Paradigms and Technological Trajectories', *Research Policy*, 11(3), pp. 147–62.

Dosi, G. (1988). 'Sources, Procedures, and Micro-Economic Effects of Innovation', *Journal of Economic Literature*, 26(3), pp. 1120–71.

Dubey, J. and Dubey, R. (2010). 'Pharmaceutical Innovation and Generic Challenge: Recent Trends and Causal Factors', *International Journal of Pharmaceutical and Healthcare Marketing*, 4(2), pp. 175–90.

Dugar, P. (2019). 'A Study of Venture Capital Firms in India'. (Unpublished doctoral dissertation), Gujarat University, Gujarat, India.

Dugar, P. and Pandit, N. (2017). 'Growth of Venture Capital and Private Equity in India', *Journal of Private Equity*, 21(1), pp. 79–93.

Dumont, B. and Holmes, P. (2002). 'The Scope of Intellectual Property Rights and their Interface with Competition Law and Policy: Divergent Paths to the Same Goal', *Economics of Innovation and New Technology*, 11(2), pp. 149–62.

Dutta, S., Lanvin, B. and Wunsch-Vincent, S. (2019). 'The Global Innovation Index 2019: Creating Healthy Lives— The Future of Medical Innovation', World Intellectual Property Organization (WIPO), Geneva, Switzerland.

Erdal, L. and Gocer I. (2015). 'The Effects of Foreign Direct Investment on R&D and Innovations: Panel Data

Analysis of Developing Countries', *Procedia—Social and Behavioural Sciences*, 195(3), pp. 749–58.

Erumban, A.A. and Das, D.K. (2016). 'Information and Communications Technology and Economic Growth in India', *Telecommunications Policy*, 40(5), pp. 412–31.

Etzkowitz, H. (2002). 'The Triple Helix of University-Industry-Government—Implications for Policy and Evaluation', *Working Paper No. 11*, Science Policy Institute, Stockholm.

Etzkowitz, H. and Leydesdorff, L. (2001). 'The Dynamics of Innovation from National Systems and Mode 2 to a Triple Helix University-Industry-Government Relationship', *Research Policy*, 29(2), pp. 109–23.

Evenson, R.E., and Westphal, L.E. (1995). 'Technological Change and Technology Strategy', *Handbook of Development Economics*, 3(A), pp. 2209–99.

Fagerberg, J. and Sapprasert, K., (2011). 'National Innovation Systems: The Emergence of a New Approach', *Science and Public Policy*, 38(9), pp. 669–79.

Falvey, R. and Foster, N. (2006). 'The Role of Intellectual Property Rights in Technology Transfer and Economic Growth: Theory and Evidence', United Nations Development Organization, Vienna.

Fernandes, A.M. and Paunov, C. (2013). 'Does Trade Stimulate Product Quality Upgrading?', *Canadian Journal of Economics/Revue Canadienne d'Économique*, 46(4), pp. 1232–64.

Forbes, N. (2017). 'India's National Innovation System: Transformed or Half Formed?', in Rakesh Mohan (Ed.), *India Transformed: 25 Years of Economic Reforms*, Penguin Random House.

Foss, N.J. and Saebi, T. (2017). 'Fifteen Years of Research on Business Model Innovation: How Far Have We Come, And

Where Should We Go? *Journal of Management*, 43(1), pp. 200–27.

Freeman, C. (1982). *The Economics of Industrial Innovation*, MIT Press, Cambridge, MA.

Freeman, C. and Perez, C. (1988). 'Structural Crises of Adjustment: Business Cycles and Investment Behaviour', in Dosi et al. (Eds) (1988), *Technical Change and Economic Theory*, Pinter Publishers, pp. 38–66.

Gaddy, B., Sivaram, V. and O'Sullivan, F. (2016). 'Venture Capital and Cleantech: the Wrong Model for Clean Energy Innovation', *MIT Energy Initiative Working Paper MITEI-WP-2016-06*. Cambridge: USA.

Garcia, R. and Calantone, R. (2002). 'A Critical Look at Technological Innovation Typology and Innovativeness Terminology: A Literature Review', *Journal of Product Innovation Management*, 19(2), pp. 110–32.

Gault, F. (2016). 'Defining and Measuring Innovation in all Sectors of the Economy: Policy Relevance', *Mimeo*, OECD Blue Sky Forum III, Ghent, Belgium.

Girma, S., Gong, Y. and Gorg, H. (2008). 'Foreign Direct Investment, Access to Finance and Innovation Activity in Chinese Enterprises', *World Bank Economic Review*, 22(2), pp. 367–82.

Globerman, S. (2012). 'Public Policies to Encourage Innovation and Productivity', Macdonald-Laurier Institute, Canada.

Gomes, C.M., Kruglianskas, I. and Scherer, F.L. (2011). 'Analysis of the Relationship between Practices of Managing External Sources of Technology Information and Indicators of Innovation Performance', *International Journal of Innovation Management*, 15(4), pp. 709–30.

González-Álvarez, N. and Nieto-Antolín, M. (2007). 'Appropriability of Innovation Results: An Empirical Study in Spanish Manufacturing Firms', *Technovation*, 27(5), pp. 280–95.

Gorodnichenko, Y., Svejnar, J. and Terrell, K. (2014). 'When Does FDI Have Positive Spillovers? Evidence from 17 Transition Market Economies', *Journal of Comparative Economics*, 42(4), pp. 954–69.

Government of India (2020). *Economic Survey 2019–2020*, Ministry of Finance, Department of Economic Affairs.

Guceri, I. (2016). 'Tax Incentives for R&D', *Mimeo*, European Tax Policy Forum.

Sachi, H. (2011). 'Building University Research and Graduate Education: Lessons from International Experience', Paper presented at a conference on Higher Education, Centre for Policy Research, New Delhi.

He, Wei. (2007). 'Examining the Determinants of R&D Investment in Developing Economies—An Empirical Study of US International R&D', Paper presented at the Doctoral Consortium, College of Business Administration, Florida International University, Miami.

Herstad, S.J., Bloch, C., Ebersberger, B. and Velde, Els van de (2010). 'National Innovation Policy and Global Open Innovation: Exploring Balances, Tradeoffs and Complementarities', *Science and Public Policy*, 37(2), pp. 113–24.

Hossain, M. (2017). 'Business Model Innovation: Past Research, Current Debates, and Future Directions', *Journal of Strategy and Management*, 10(3), pp. 342–59.

Houben, A. and Kakes, J. (2002). 'ICT Innovation and Economic Performance: The Role of Financial Intermediation', *Kyklos*, 55(4), pp. 543–62.

Humphrey, J. and Schmitz, H. (2002). 'How Does Insertion in Global Value Chains Affect Upgrading in Industrial Clusters?', *Regional Studies*, 36(9), pp. 1017–27.

Hurmelinna, P., Kyläheiko, K. and Jauhiainen, T. (2007). 'The Janus Face of the Appropriability Regime in the Protection of Innovations: Theoretical Re-Appraisal and Empirical Analysis', *Technovation*, 27(3), pp. 133–44.

Iammarino, S. and McCann, P. (2006). 'The Structure and Evolution of Industrial Clusters: Transactions, Technology and Knowledge Spillovers', *Research Policy*, 35(7), pp. 1018–36.

IEA (2003). *Renewables for Power Generation: Status and Prospects*, Paris: OECD Publishing.

Inglesi-Lotz, R. (2016). 'The Impact of Renewable Energy Consumption to Economic Growth: A Panel Data Application', *Energy Economics*, 53, pp. 58–63.

Iwasaki, I. and Tokunaga, M. (2016). 'Technology Transfer and Spillovers from FDI in Transition Economies: A Meta-Analysis', *Journal of Comparative Economics*, 44(4), pp. 1086–114.

Jayal, N.G. (2020). 'NEP 2020 on Higher Education', *India Forum*, https://www.theindiaforum.in/article/nep-2020-higher-education

Jacobides, M.G., Knudsen, T. and Augier, M. (2006). 'Benefiting from Innovation: Value Creation, Value Appropriation and the Role of Industry Architectures', *Research Policy*, 35(8), pp. 1200–21.

Jensen, M.B., et al. (2007). 'Forms of Knowledge and Modes of Innovation', *The Learning Economy and the Economics of Hope*.

Johansson, B. and Loof, H. (2005). 'FDI Inflows to Sweden: Consequences for Innovation and Renewal', *Working Paper Series No. 30*, Economics and Institutions of Innovation, Royal Institute of Technology, CESIS—Centre for Excellence in Innovation Studies.

Kanwar, S. (2013). 'Innovation, Productivity and IPRs', *Working Papers No 230*, Centre for Development Economics, Delhi School of Economics, Delhi University.

Kapur, D. and Mehta, P.B. (Eds). (2017). *Navigating the Labyrinth: Perspectives on India's Higher Education*, Orient Blackswan.

Kathuria, V. (2008). 'The Impact of FDI Inflows on R&D Investment by Medium- and High-Tech Firms in India in the Post-Reform Period', *Transnational Corporations*, 17(2), p. 45.

Kathuria, V. (2010). 'Does the Technology Gap Influence Spillovers? A Post-Liberalization Analysis of Indian Manufacturing Industries', *Oxford Development Studies*, 38(2), pp. 145–70.

Kerr, W.R. and Nanda, R. (2015). 'Financing Innovation', *Annual Review of Financial Economics*, 7, pp. 445–62.

Kline, S.J. and Rosenberg, N. (1986). 'An Overview of Innovation', in R. Landau and N. Rosenberg (Eds), *The Positive Sum Strategy: Harnessing Technology for Economic Growth*, Washington DC: National Academy Press.

Kodama, Fumio, Kano, Shingo and Suzuki, J. (2007). 'Beyond Absorptive Capacity: The Management of Technology for a Proactive Corporate Strategy towards University-Industry Links', In Yusuf and Nabeshima (Eds).

Kokko, A., Tansini, R. and Zejan, M.C. (1996). 'Local Technological Capability and Productivity Spillovers from FDI in the Uruguayan Manufacturing Sector', *Journal of Development Studies*, 32(4), pp. 602–11.

Krishna, K.L. et al. (2018). 'ICT Investment and Economic Growth in India: An Industry Perspective', *Working Paper No. 284*, Centre for Development Economics.

Krishna, R. (Ed.). (2012). *Applications of Pharmacokinetic Principles in Drug Development*, Springer Science and Business Media.

Krishna, V.V. and Bhattacharya, S. (2009). 'Internationalization of R&D and Global Nature of Innovation: Emerging Trends in India', *Working Paper No. 123*, Asia Research Institute, National University of Singapore.

Krishna, V. and Patra, S. (2015). 'Research and Innovation in Universities in India', in Varghese, N.V. and Malik, G. (Eds), *India Higher Education Report*, New York: Routledge.

Krishnan, R.T. (2012). 'Innovation Strategies of Indian Market Leaders', *Journal of Indian Business Research*, 4(2), pp. 92–96.

Lall, S. (1994). 'The East Asian Miracle: Does the Bell Toll for Industrial Strategy?', *World Development*, 22(4), pp. 645–54.

Lan, Xue. and Liang, Zheng. (2006). 'Multinational R&D in China: Myth and Realities', School of Public Policy and Management, Tshingua University, Beijing.

Leiponen, A. and Byma, J. (2009). 'If You Cannot Block, You Better Run: Small Firms, Cooperative Innovation, and Appropriation Strategies', *Research Policy*, 38(9), pp. 1478–88.

Lerner, J. and Tag, J. (2013). 'Institutions and Venture Capital', *Industrial and Corporate Change*, 22(1), pp. 153–82.

Lipsey, R.G. (2002). 'Some Implications of Endogenous Technological Change for Technology Policies in Developing Countries', *Economics of Innovation and New Technology*, 11(4-5), pp. 321–51.

Lockett, A., Siegel, D., Wright, M. and Ensley, M.D. (2005). 'The Creation of Spin-Off Firms at Public Research

Institutions: Managerial and Policy Implications', *Research Policy*, 34(7), pp. 981–93.

Lundvall, B.Å. (2007). 'Post Script: Innovation System Research—Where It Came From and Where It Might Go', The Learning Economy and the Economics of Hope.

Malen, J. and Marcus, A.A. (2017). 'Promoting Clean Energy Technology Entrepreneurship: The Role of External Context', *Energy Policy*, 102, pp. 7–15.

Mallapur, C. (2020). 'High Targets and Wasted Funds: The Problems with the Skill India Programme, https://scroll.in/article/844871/high-targets-and-wasted-funds-the-problems-in-the-skill-india programme

Mani, S. (2009). 'Is India Becoming More Innovative Since 1991? Some Disquieting Features', *Economic and Political Weekly*, 44(46), pp. 41–51.

Mani, S. (2017). 'Robot Apocalypse: Does It Matter for India's Manufacturing Industry?', *Working Paper No. 474,* Centre for Development Studies.

Mani, S. (2018). 'What Is Happening to India's R&D Funding?', *Economic and Political Weekly*, 53(14), pp. 12–14.

Mani, S. (2020). 'India's Quest for Technological Self-Reliance', *Working Paper No. 496,* Centre for Development Studies.

Mani, S. and Nabar, J. (2016). 'Is the Indian Government Justified in Lowering Tax Incentives for R&D Expenditure of Firms?', *Economic and Political Weekly*, 2, p. 30.

Mann, P., Nayak, S. and Aggarwal, V. (2015). 'India's Manufacturing Exports: Technology Intensity Transition', *International Research Journal of Social Sciences*, 4, pp. 67–75.

Mansfield, E. et al. (Eds) (1971). 'Innovation and Discovery in the Ethical Pharmaceutical Industry', *Research and*

Innovation in the Modern Corporation, London: Palgrave Macmillan.

Martin, S. and Scott, J.T. (2000). 'The Nature of Innovation Market Failure and the Design of Public Support for Private Innovation', *Research Policy*, 29(4-5), pp. 437–47.

Mazzucato, M. (2018). 'Mission-Oriented Innovation Policies: Challenges And Opportunities', *Industrial and Corporate Change*, 27(5), pp. 803–15.

McAdam, M., Miller, K. and McAdam, R. (2016). 'Situated Regional University Incubation: A Multi-Level Stakeholder Perspective', *Technovation*, 50–51, pp. 69–78.

Merlevede, B., Schoors, K. and Spatareanu, M. (2014). 'FDI Spillovers and Time Since Foreign Entry', *World Development*, 56, pp. 108–26.

Metcalfe, S. (1985). 'On the Diffusion of Innovation and the Evolution of Technology', In William, B.R. and J.A. Bryan Brown (Eds), *Knowns and Unknowns of Technical Change*, London: The Technical Change Centre.

Metcalfe, S. and Ramlogan, R. (2008). 'Innovation Systems and the Competitive Process in Developing Economies', *Quarterly Review of Economics and Finance*, 48(2), pp. 433–46.

Meyer, K.E. (2003). 'FDI Spillovers in Emerging Markets: A Literature Review and New Perspectives', *DRC Working Paper No. 15,* Centre for New and Emerging Markets, London Business School.

Mian, S., Lamine, W. and Fayolle, A. (2016). 'Technology Business Incubation: An Overview of the State of Knowledge', *Technovation*, 50–51, pp. 1–12.

Micaëlli, J.P., Forest, J., Coatanéa, E. and Medyna, G. (2014). 'How to Improve Kline and Rosenberg's Chain-Linked Model of Innovation: Building Blocks and Diagram-

Based Languages', *Journal of Innovation Economics Management*, (3), pp. 59–77.

Miravete, E.J. and Pernias, J.C. (2006). 'Innovation Complementarity and Scale of Production', *Journal of Industrial Economics*, 54(1), pp. 1–29.

Mishra, P. (2011). 'Determinants of Inter-Industry Variations, Mergers and Acquisitions: Empirical Evidence from Indian Manufacturing Sector', *Artha Vijnana*, LIII (1), pp. 1–22.

Mohnen, O. and Roller, L.H. (2005). 'Complementarities in Innovation Policy', *European Economic Review*, 49(6), pp. 1431–50.

Morris, S. (1997). 'Why Not Push for 9 per cent Growth?', *IIMA Working Paper No. 1364*, Indian Institute of Management Ahmedabad.

Mowery, D. and Rosenberg, N. (1979). 'The Influence of Market Demand upon Innovation: A Critical Review of Some Recent Empirical Studies', *Research Policy*, 8(2), pp. 102–53.

Mowery, D. and Sampat, B. (2005). *Universities and Innovation*, The Oxford Handbook on Innovation.

Mowery, D.C., Nelson, R., Sampat, B. and Ziedonis, A. (2001). 'The Growth of Patenting and Licensing by US Universities: An Assessment of the Effects of the Bayh–Dole Act of 1980', *Research Policy*, 30(1), pp. 99–119.

Mowery, D.C., and Sampat, B.N. (2005). *Universities and Innovation, The Oxford Handbook on Innovation*.

Nabar, J. (2018). 'Strengthening India's Innovation System in CTIER (2018)', *CTIER Handbook: Technology and Innovation in India 2019*, Centre for Technology Innovation and Economic Research, Pune, pp. 12–22.

Nair, A., Guldiken, O., Fainshmidt, S. and Pezeshkan, A. (2015). 'Innovation in India: A Review of Past

Research and Future Directions', *Asia Pacific Journal of Management*, 32(4), pp. 925–58.

Nepal, R. (2012). 'Roles and Potentials of Renewable Energy in Less-Developed Economies: The Case of Nepal', *Renewable and Sustainable Energy Reviews*, 16(4), pp. 2200–06.

Nga N. Ho-Dac. (2020). 'The Value of Online User Generated Content in Product Development', *Journal of Business Research*, 112, pp. 136–46.

Nonaka, I. and Takeuchi, H. (1995). *The Knowledge-Creating Company: How Japanese Companies Create the Dynamics of Innovation*, Oxford University Press.

O'Sullivan, M. (2005). 'Finance and Innovation', in J. Fagerberg, D. Mowery and R.R. Nelson (Eds), *The Oxford Handbook of Innovation*, Oxford: Oxford University Press.

OECD (2011). '*Measuring Innovation: A New Perspective*', https://www.oecd.org/site/innovationstrategy/measuringi nnovationanewperspective-onlineversion.htm

Ordover, J.A. (1991). 'A Patent System for Both Diffusion and Exclusion', *Journal of Economic Perspectives*, 5(1), pp. 43–60.

Owen, A.D. (2006). 'Renewable Energy: Externality Costs as Market Barriers', *Energy Policy*, 34(4), pp. 632–42.

Parrilli, M.D. and Heras, H.A. (2016). 'STI and DUI Innovation Modes: Scientific-Technological and Context-Specific Nuances', *Research Policy*, 45(4), pp. 747–56.

Perkmann, M., et al. (2013). 'Academic Engagement and Commercialisation: A Review of the Literature on University–Industry Relations', *Research Policy*, 42(2), pp. 423–42.

Ramdani, B., Binsaif, A. and Boukrami, E. (2019). 'Business Model Innovation: A Review and Research Agenda', New England Journal of Entrepreneurship, 22(2), pp. 89–108.

Rao, K.S. and Dhar, B. (2018). 'India's Recent Inward Foreign Direct Investment: An Assessment', *MPRA Paper No. 88992*, Institute for Studies in Industrial Development.

Reddy, K.S., Xie, E. and Tang, Q. (2016). 'Higher Education, High-Impact Research, and World University Rankings: A Case of India and Comparison with China', *Pacific Science Review B: Humanities and Social Sciences*, 2(1), pp. 1–21.

Reitzig, M. (2004). 'The Private Values of "Thickets" and "Fences": Towards an Updated Picture of the Use of Patents Across Industries', *Economics of Innovation and New Technology*, 13(5), pp. 457–76.

Rosenberg, N. (1982). *Inside the Black Box: Technology and Economics*, Cambridge University Press.

Rothaermal, F.T. and Hess, A.M. (2007). 'Building Dynamic Capabilities: Innovation Driven by Individual, Firm and Network Level Effects', *Organization Science*, 18(6), pp. 898–921.

Sabarinathan, G. (2019). 'Angel Investments in India—Trends, Prospects and Issues', *IIMB Management Review*, 31(2), pp. 200–14.

Scherer, F.M., and Ross, D. (1990). 'Industrial Market Structure and Economic Performance', University of Illinois at Urbana-Champaign's Academy for entrepreneurial leadership, historical research reference in entrepreneurship.

Sharma, S. and Vohra, N. (2020). 'Incubation in India–A Multilevel Analysis', *IIMA Working Paper No. 2020-03-01*, Indian Institute of Technology Ahmedabad.

Shin, D.H., Kim, H. and Hwang, J. (2015). 'Standardization Revisited: A Critical Literature Review on Standards and Innovation', *Computer Standards and Interfaces*, 38, pp. 152–57.

Shoaib, A. and Ariaratnam, S. (2016). 'A Study of Socio-economic Impacts of Renewable Energy Projects in Afghanistan', *Procedia Engineering*, 145, pp. 995–1003.

Sloan, B. (2006). 'Developing the Linkage between Policy and Innovation Measurement', in Blankley, W., M. Scerri and N. Molotja (Eds), *Measuring Innovation in OECD and Non-OECD Countries: Selected Seminar Papers*, Human Science Research Council: pp. 43–58.

So, A.D. et al. (2008). 'Is Bayh-Dole Good for Developing Countries? Lessons from the US Experience', *PloS Biol*, 6, p. 262.

Song, H. and Vandenbussche, H. (2008). 'Trade Policy and Innovation', *LICOS Discussion Paper, No. 200*, Katholieke Universiteit Leuven, LICOS Centre for Institutions and Economic Performance, Leuven.

Steinmueller, W.E. (2010). 'Economics of Technology Policy', *Handbook of the Economics of Innovation*, 2, pp. 1181–1218.

Stiebale, J. and Reize, F. (2008). 'The Impact of FDI on Innovation in Target Firms', *Ruhr Economic Paper No. 50*, Rheinisch Westfälisches Institut für Wirtschaftsforschung (RWI), Essen.

Stoneman, P. and Diederen, P. (1994). 'Technology Diffusion and Public Policy', *Economic Journal*, 104(425), pp. 918–30.

Strube, E. and Resende, M. (2009). 'Complementarity of Innovation Policies in the Brazilian Industry: An Econometric Study', *CESifo Working Paper No. 2780*, Centre for Economic Studies and ifo Institute (CESifo), Munich.

Teece, D.J. (1986). 'Profiting from Technological Innovation: Implications for Integration, Collaboration, Licensing and Public Policy', *Research Policy*, 15, pp. 285–305.

Thomä, J. and Bizer, K. (2013). 'To Protect or Not to Protect? Modes of Appropriability in the Small Enterprise Sector', *Research Policy*, 42(1), pp. 35–49.

von Hippel, E. (2009). 'Democratizing Innovation: The Evolving Phenomenon of User Innovation', *International Journal of Innovation Science*, 1(1), pp. 29–40.

Wang, C.C. and A, Wu. (2016). 'Geographical FDI Knowledge Spillover and Innovation of Indigenous Firms in China', *International Business Review*, 25(4), pp. 895–906.

Wessner, C.W. (Ed.) (2008). 'An Assessment of the SBIR Program', National Research Council (US) of the National Academics, Washington (DC): National Academies Press (US).

Wellhausen, R.L. (2013). 'Innovation in Tow: R&D, FDI and Investment Incentives', *Business and Politics*, 15(4), pp. 467–91.

Westmore, B. (2013). 'Innovation and Growth: Considerations for Public Policy', *Review of Economics and Institutions*, 4(3), p. 50.

Williams, H.L. (2016). 'Intellectual Property Rights and Innovation: Evidence from Health Care Markets', *Innovation Policy and the Economy*, 16(1), pp. 53–87.

Wolf, A. (2002). *Does Education Matter?: Myths about Education and Economic Growth*, Penguin UK.

Wu, Y., Popp, D. and Bretschneider, S. (2007). 'The Effects of Innovation Policies on Business R&D: A Cross-National Empirical Study', *Economics of Innovation and New Technology*, 16(4), pp. 237–53.

Yarnell, A. (2006). 'Indian Science Rising', *Chemical and Engineering News*, 84(2), pp. 12–20.

Youtie, J. and Shapiro, P. (2008). 'Building an Innovation Hub: A Case Study of the Transformation of University

Roles in Regional Technological and Economic Development', *Research Policy*, 37(8), pp. 1188–204.

Yusuf, S. (2007). 'University-Industry Links: Policy Dimensions', in Yusuf, S. and Nabeshima, K. (Eds) (2007).

Yusuf, S., Nabeshima, K. and Yamashita, S. (2008). 'Growing Industrial Clusters in Asia: Serendipity and Science. Directions in Development: Private Sector Development', Washington DC: World Bank.

Yusuf, S., and Nabeshima, K. (2007). *How Universities Promote Economic Growth*, World Bank.

Zhang, X. and Wu, J. (2014). 'Research on Effectiveness of the Government R&D Subsidies: Evidence from Large and Medium Enterprises in China', *American Journal of Industrial and Business Management*, 4(9), p. 503.

Zhao, M. (2006). 'Conducting R&D in Countries with Weak Intellectual Property Rights Protection', *Management Science*, 52(8), pp. 1185–99.

Zott, C., Amit, R. and Massa, L. (2011). 'The Business Model: Recent Developments and Future Research', *Journal of Management*, 37(4), pp. 1019–42.